Machine Woodworking Technology

For Hand Woodworkers

To my wife and family, without whose co-operation I would not have had the time to write.

Machine Woodworking Technology

For Hand Woodworkers

F. E. Sherlock
Senior Lecturer in Charge of Furniture and Machine Woodworking Courses,
Tottenham College of Technology

Butterworths
London - Boston - Sydney - Wellington - Durban - Toronto

First published 1973
reprinted 1977, 1979, 1982, 1985

ISBN 0 408 00113 5

Printed in England by Page Bros (Norwich) Ltd

Preface

Today, crafts are tending to overlap, and this is particularly true in the case of woodworking. Every syllabus of study for a woodworking examination contains elements of machine woodworking, and joiners are required to list machine sequences, to mark out for machine work, to explain cutting principles and to demonstrate a knowledge of machine technology.

Students of the 555/425 range of courses now combine at the first year entry to study both hand and machine work in wood, metals and plastic. In Stage 2 of these courses students on furniture courses must be able to demonstrate a knowledge of machine woodworking in order to obtain passes in two sections of the examinations. Vehicle body students, shop fitters and sawmillers are also required to have the same knowledge and to pass similar examinations; their most important need is to know about machines.

This book was planned and written to cover such a broad field of technology that it would be suitable as a textbook for all hand woodworkers. However, it soon became apparent that it would also cover the needs of machine woodworking students, and for this reason more basic technology was added.

By the time the book is published, metrication and the introduction of S.I. units will be well under way. All examples of calculations are worked in this medium and expressed in a manner that I have found my own students to understand and assimilate. 'The wish and the will' behind the effort is that machines will be used with greater skill and understanding, and with greater profit, once their raison d'être is understood.

ACKNOWLEDGEMENTS

Without the assistance of many friends and colleagues much of the material present would not have been available. Very especially, the manuscript was typed by my friend Audrey Caplin, and read and scrutinised by my colleague Peter Saggers.

Technical information relating to their own products was supplied by Mr. S. N. M. Evans of Rye Machinery Sales Ltd., and by Mr. M. J. McVey of Dexter and Co., High Wycombe. Interwood Ltd. of London and Wadkin Ltd. of Leicester both freely provided machine literature. To all of these, I express my profound gratitude.

SAFETY

It should always be remembered that machine woodworking is inherently dangerous. However, skilful and intelligent use of the machines will obviate many of the risks, and strict adherence to the Factories Acts will minimise the remainder.

The Factories Acts and the Machine Woodworking Regulations form part of the syllabuses of the relevant City and Guilds courses, and since 1970 the Woodworking Machinery Regulations have been studied and amended in draft form by interested parties. The first draft has now been amended again and the second proposed draft was published in the Autumn of 1972. Subject to them achieving the status of law, the new regulations will be available for study and application sometime in 1973. Examination syllabuses will then include the current regulations, but students sitting examinations in 1973 and 1974 should continue to study the earlier regulations. These are The Woodworking Machinery Regulations, 1922 and Amendments of 1927 and 1945.

F.E.S.

Contents

1 Machine Sawing

The saw as a tool

If a hand saw is to be used to produce flat, clean surfaces, the saw must be in good order and it must be used with skill. The same considerations govern the use of a circular saw, although the emphasis of skill is biased towards the preconditioning of the saw rather than in the manual application of wood to the saw. Nevertheless, circular saws are tools which must be used skilfully if acceptable sawing is to be achieved.

An understanding of the fundamentals of circular saws and of the machines on which they are used is essential if the operator is to become skilled in their use. Hand woodworkers have a tendency to accept their saws as tools which are perfect, and they rarely consider the design and preparation of them; this is acceptable as long as the saw is in good condition. However, machine saws cannot be treated in this way and a full basic understanding of them is essential.

The circular saw

CUTTING ACTION

A circular saw is a steel disc having around its periphery a number of shaped teeth. The shape of these teeth can vary considerably, but generally they fall into five main groups, each best suited to a particular class of work. These five groups are ripsawing, cross-cutting, dimension sawing, fine sawing (such as plywood cutting) and abrasive sawing (requiring carbide-tipped saw teeth). Figure 1.1 shows the general arrangement and shape of these five groups, and it is worthwhile remembering that for the purpose of answering City & Guilds examination questions it is sufficient to show three teeth only.

Saw teeth have different shapes to enable the saw to cut most efficiently on a narrow range of cutting, and no tooth will cut at its best on every type of work. This is because wood is fibrous; wood cut along its grain direction reacts in a different manner to wood cut across its grain. Wood fibres cling together edge on edge, and to cut a board end to end requires the physical removal of a narrow band of wood, thus separating one board into two. The narrow band has a width known as the saw *kerf* which is the width of the saw cut, and through the length of the board this strip must be removed chip by chip, by tooth following

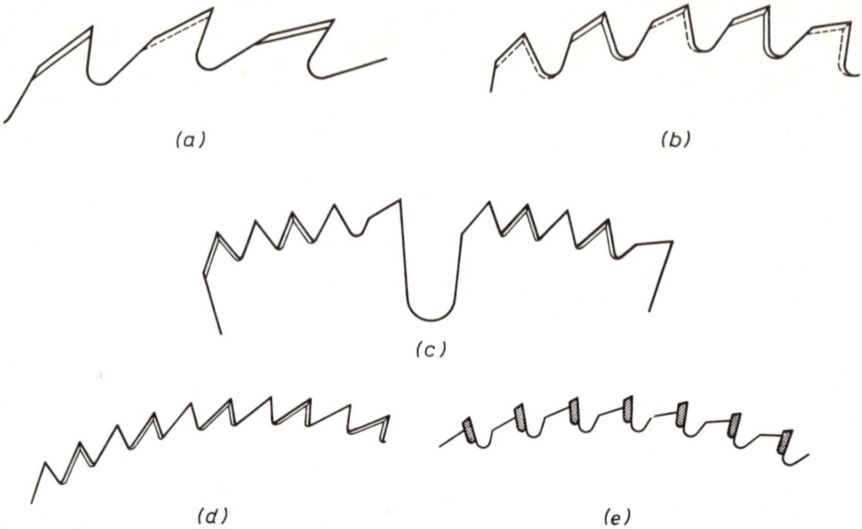

Figure 1.1 *Saw teeth.* (a) *Ripsawing teeth, which lean forward, are said to have a positive hook or rake,* (b) *cross-cutting teeth, which lean back, are said to have a negative hook,* (c) *dimension sawing requires a collection of rip and cross-cut teeth,* (d) *plywood and other laminated boards require numerous small teeth to avoid whiskered edges and* (e) *carbide-tipped saws which are used for abrasive cutting conditions*

tooth. The saw tooth for this job must present a chisel edge to the wood fibres, chopping them off cleanly in length and severing them cleanly on side.

When cutting across, or at an angle to the grain, the wood fibres are presented to the saw teeth on edge rather than on end as in ripsawing (end to end of board), and rather than separating cleanly they tend to break and show their fibrous nature by producing whiskers on the underside of the wood being cut. To overcome this, saw teeth for cross-cutting require sharp points rather than rip-teeth chisel edges. In addition, the tooth points must trail into each fibre and not dig into them. One half of the teeth of a cross-cut saw sever fibres to the right of the saw plate and one half sever fibres to the left, and this takes place over the narrow width of the saw kerf. The short length of fibre left unsupported between left and right points falls away easily. The fact that the fibres are severed with needle points cuts down the incidences of whiskers to a minimum and leaves an acceptable clean cut edge.

Basically this covers the reasons for the varying designs of saw teeth for wood, and two basic forms are established, viz. ripsaw teeth that must have chisel edges and must incline towards the wood, and cross-cut saw teeth that must have needle points and incline away from the wood. As woodcutting machinists put it, 'Ripsaw teeth have a positive hook or rake, and cross-cut teeth have a negative hook or rake'. With these principles in mind it can be seen that all circular saw teeth must be designed and maintained to do a particular job of cutting.

Change of tooth shape is made necessary by variation in the cutting job. Plywood, being laminated veneers set at right angles to each other and requires neutral angle teeth that are very fine, as the lower face is always a thin veneer and

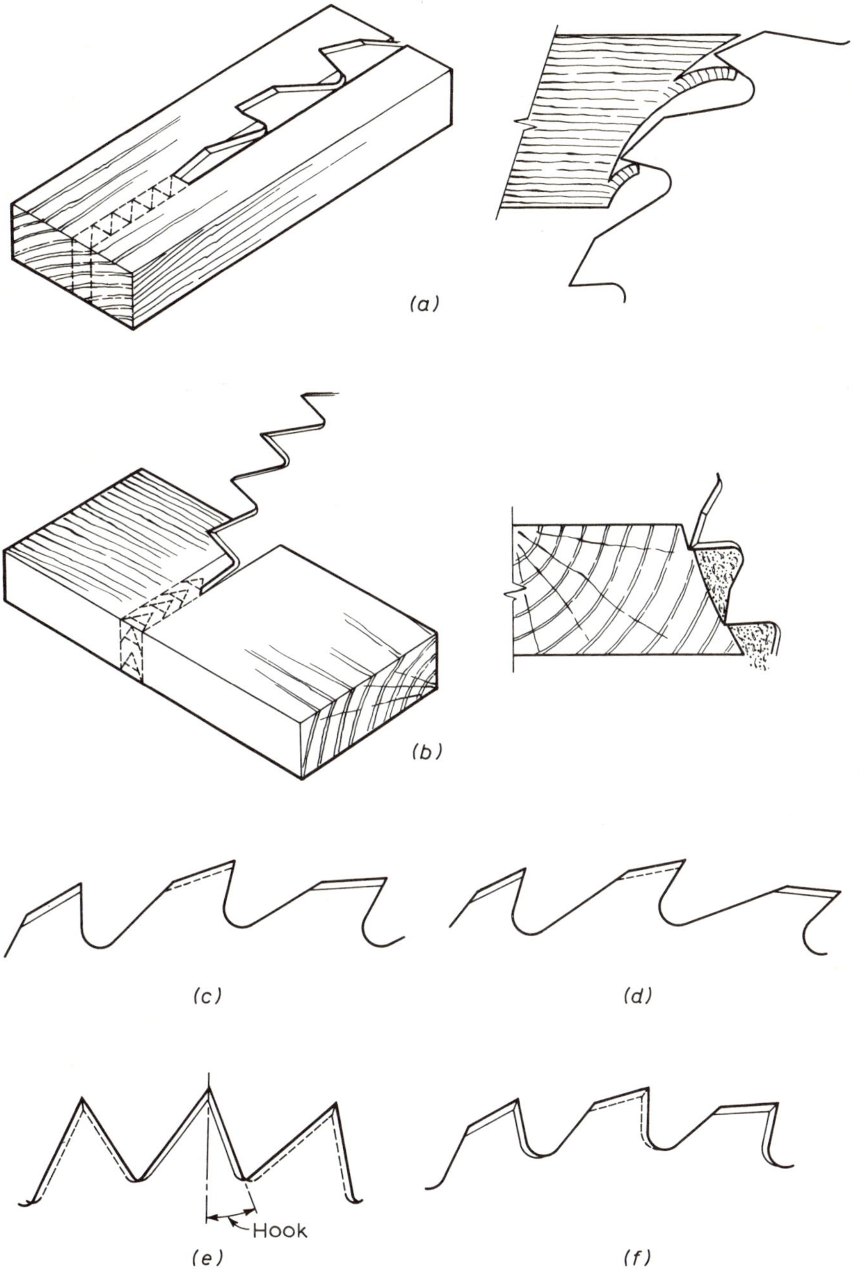

Figure 1.2 Ripsawing and cross-cutting, and hardwood and softwood teeth. (a) *Ripsawing: chip by chip removal of stock,* (b) *cross-cutting: needle teeth cut cleanly at each side,* (c) *strong hardwood rip teeth with a low hook angle,* (d) *softwood rip teeth: sharp chisel teeth with high hook angle and plenty of clearance,* (e) *softwood cross-cut teeth: needle-point teeth with the maximum clearance and* (f) *hardwood cross-cut teeth: needle-point teeth strongly backed up*

easily split or 'whiskered'. Dimension sawing, where the wood is cut at any angle in any grain direction, needs a combination of rip teeth and cross-cut teeth, and the group shown in Figure 1.1*c* has one rip and four cross-cut teeth in groups around the saw periphery. The cross-cut teeth project 0·5 mm beyond the chisel edge of the rip teeth.

Tooth design must also cater for variation in wood structure. Hardwood boards require tougher, more rigid teeth than softwood boards, and to this end hardwood teeth are more square and have less hook than softwood teeth. Hardwood fibres, being more inclined to split and splinter, require more cutting teeth than softwood fibres, and so it is that hardwood cross-cut saws have smaller teeth than those for softwood.

Figure 1.2 shows the approach of two types of saw teeth to their job of cutting, and also shows variation in tooth form within particular saw groups.

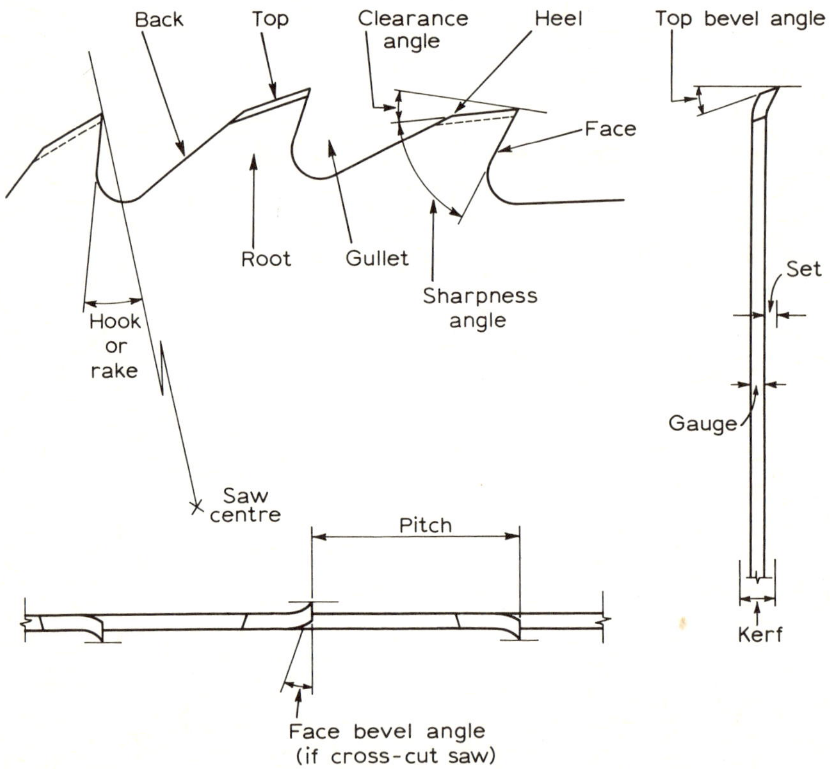

Figure 1.3 Saw terms

Tungsten-carbide-tipped (T.C.T.) saw teeth incorporate all of these principles. Tungsten carbide is a conglomeration of hard tungsten carbide particles bonded in a matrix of cobalt*. It is used because of its extremely high wear

*Description from the literature of *Protolite Ltd.*, makers of tungsien carbide.

resistance. Each time a steel saw is sharpened it is reduced in diameter and is non-productive for its sharpening period. Tungsten carbide cuts for many times longer than standard saw steel. Figure 1.3 shows the names given to the various parts and angles contained in the design of saw teeth.

SAW TYPES

Saws are made from carbon steel alloyed with other materials to give greater wear resistance. The saw plate has a thickness known as the *gauge* and is standard for each diameter, although saws may be obtained up to two gauges thinner or thicker than standard.

T.C.T. saws tend to have a thicker plate. Since the tips are hard and brittle undue vibration leads to loss of sharp points and vibration in cut will negate the engineered accuracy of the saws. Saw plates are produced to suit varying work types. For straightforward rip, or cross-cutting, the plate is parallel in thickness and the saw is called a *plate* saw. For dimension sawing, and for saws designed to leave a planed surface, the plate is *hollow ground* which means that the teeth of the saw are thicker than the portion of the saw plate adjacent to the centre collar area, which is again thick for strength.

The thickest part of the saw plate is at the extreme rim. The grinding that forms the hollow may be either a double or a single hollow grind. With a double grind the teeth are almost parallel, and then, at the root of the teeth, the grinding angle increases, and this gives strong points to the teeth and very smooth cutting at the sides.

To save wood when ripsawing large quantities of moulding stock or fence boards, saws are used which are ground very thin towards the rim. These saws

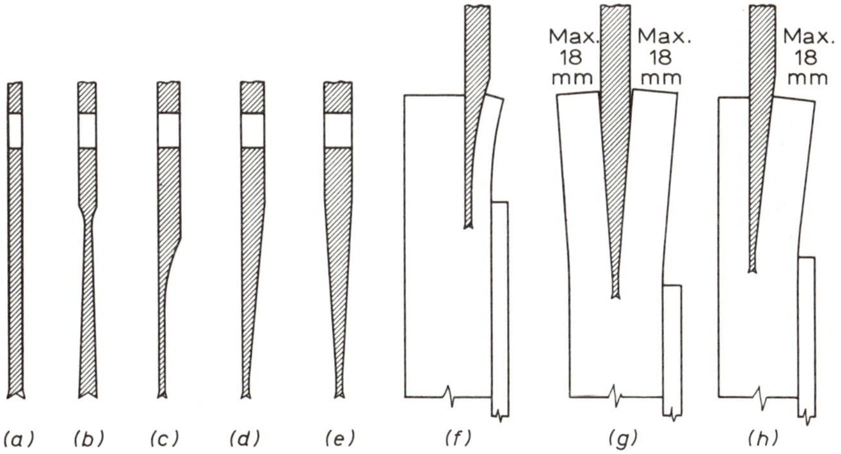

Figure 1.4 Saw plates and their uses. (a) *Plate saw,* (b) *hollow-ground saw,* (c) *ground-off saw,* (d) *swage saw,* (e) *taper saw,* (f) *ground-off saw, which is limited to the cutting of very thin strips or boards,* (g) *taper saw that can only cut thin boards down the centre with each cut board limited to approximately 18 mm and* (h) *swage saw which is limited to cutting strips of 18 mm maximum*

are called *thin rim* saws and are of three types, the *swage*, the *taper* saw and the *ground-off* saw, each being tapered to a different shape. Their purpose is the same, i.e. to save wood, but each has its own range of work.

Figure 1.4 shows sections through each of these saws and indicates the application of thin rim saws.

TREATMENT OF SAW PLATES

Saw plates require considerable pretreatment prior to being put to work. Assuming that they have been received in good condition and are fit for their intended purpose, the sawyer has the job of keeping them in good condition. To do this he will need the assistance of a *saw doctor* unless he attends to his own saws. Most often today the saws go out to an agency, or are channelled into a special saw shop in the factory.

The treatment required for saws is four-fold. The saw side clearance must be maintained. The saw teeth must be kept in a true cutting circle, and the plate must be kept flat and in the correct state of tension.

Saws are sharpened by hand with mill-saw files, which are flat with round edges and have teeth cut to a grade known as 'second cut, single'. The saw is fitted to a saw horse and the teeth are filed by hand to the correct shape and angle. For large scale sharpening and for remedial treatment to badly shaped teeth, automatic grinders are used, but most saw doctors feel that hand filing produces a sweeter-cutting saw. Teeth can become badly shaped as a result of unskilled hand filing.

Side clearance to the saw plate is given by 'setting' the teeth. Alternate teeth are sprung to one side using a 'gate' saw set, then the remaining teeth are sprung to the other side, thus leaving the saw with half of its teeth set to clear one side and the other half set to clear the other side. This procedure produces a kerf wider than the saw plate thickness and is exactly the same in principle as the process used in hand saws, although a gate-saw set is used together with a gauge for control of the amount of set. The kerf width is the width of cut and is the saw gauge thickness plus twice the amount of set. The amount of set required varies with each saw and job; woolly, wet timber requires more set than hard, dry timber, which will vary then from 0·3 mm to 0·4 mm.

To keep the saw round it is 'stoned'. This means that whilst the saw is running, a piece of abrasive stone is brought lightly into contact with its rotating periphery. The cut is then tested for squareness, and the teeth inspected to see that most of them have been in contact with the stone. If this is so, the saw is sharpened using the stone marks as a guide to the amount to remove with a file.

This procedure would not in itself keep the saw round. The centre hole of the saw may not be an accurate fit to the saw mandrel, with the result that the saw will fall into its lowest position, and the teeth at the lowest part of the saw will be further away from the axis of rotation. If stoned in this position the teeth at the lowest part of the saw will receive more heavy stoning than those at the topmost part, if indeed these are stoned at all. If filed in this condition the saw will only cut as round if it is put back to the mandrel in exactly the same position as that in which it was stoned.

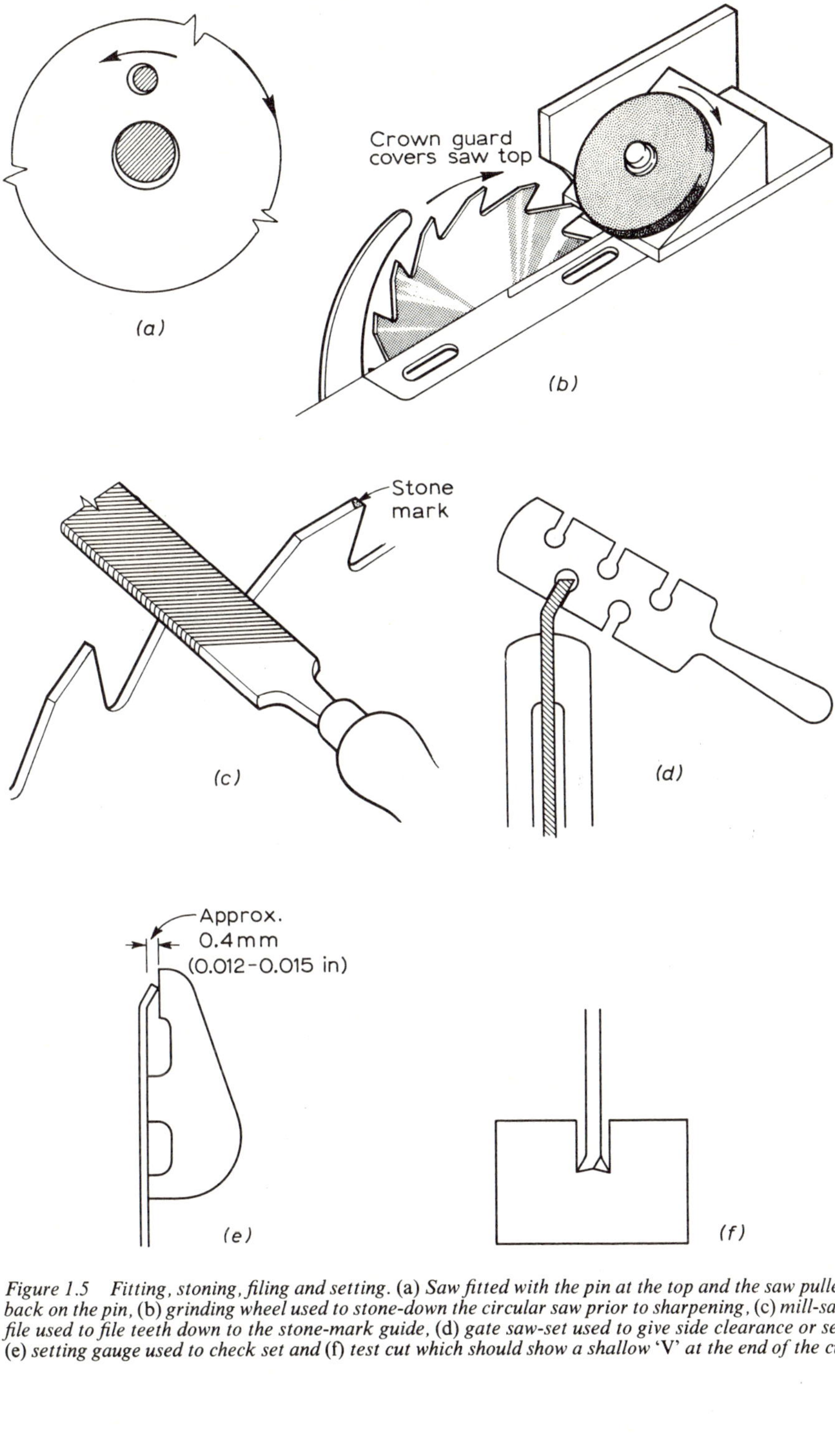

Figure 1.5 *Fitting, stoning, filing and setting*. (a) *Saw fitted with the pin at the top and the saw pulled back on the pin*, (b) *grinding wheel used to stone-down the circular saw prior to sharpening*, (c) *mill-saw file used to file teeth down to the stone-mark guide*, (d) *gate saw-set used to give side clearance or set*, (e) *setting gauge used to check set and* (f) *test cut which should show a shallow 'V' at the end of the cut*

To overcome this possible source of malfunction in cutting, the saw is always put to the mandrel in the same position. This is with the drive pin at the top dead centre and the saw pulled back into contact with it. This method ensures that at all times the maximum number of teeth are kept running truly concentric to the axis of rotation. Figure 1.5 illustrates setting, stoning and sharpening.

TENSIONING THE SAW

All the aforementioned facets of saw condition may be, and are, subject to variation by sawyers and saw doctors. Each man will have his own ideas relating to correct tooth shape, correct set and correct type of saw for each job. There is one thing on which most will agree, i.e. the need to run a saw at its correct cutting speed, which will vary for each saw type, saw diameter and type of machine.

The cutting speed of a circular saw has always been stated as rim or peripheral speed in terms of feet per minute. Obviously then, for a given saw diameter, there must be a given number of revolutions per minute. As the saw is sharpened it loses diameter and as no saw benches are built with variable speed drives, the rim speed will fall below the optimum for cutting wood, which is 10 000 ft/min or 3048 m/min. Using S.I. units this should be expressed as metres per second. 3048 m/min gives 50·8 m/s, but for convenience this can be rounded off to 50 m/s, which is therefore the figure for standard plate saws. Thin rim saws run faster, at 60 m/s. Saws of under approximately 450 mm diameter will tend to be of a heavier standard gauge, and, according to the machine on which they run, rotate faster to give up to 72 m/s. These figures are intended to give only a good working rule. When we go completely over to S.I. units a new standard will almost certainly be established.

It is the rim speed that governs the amount of tension required in a saw. The thinner the saw, the greater the tension it will require with the reverse applying to thicker saws. Faster saws require more tension than slower ones but all must be tensioned to suit their speed, diameter and gauge.

Tension is a tensile stress. In a saw it is the outward pull of the teeth due to centrifugal force that sets up a tensile stress in the saw plate, and thus, when running, the saw plate is in tension from collar to rim. A further cause of tension in the saw plate is the expansion of the saw rim due to the frictional heat generated in cutting. It is not technically correct to speak, as saw doctors do, of ‘Putting the tension into a saw’. It is possible to pre-stress the saw so that the correct internal tension will be created when the saw is running and cutting at its correct operational speed, and for want of a better expression it is usual to speak of this pre-stressing as ‘tensioning’.

A flat steel disc, if perfectly flat, will remain flat and true if run at low speed on a mandrel. When variable stresses are created in the disc, due to the braking effect of sawing (if the disc is a saw), to the heating effect of friction (in sawing) and to the outward pull of centrifugal force if the revolutions per second are increased, the rim area will expand. If the whole disc expands at an even rate it will remain flat. As it is in practice however with circular saws, the collar area clamped between drive collars does not expand and the rim expands more in

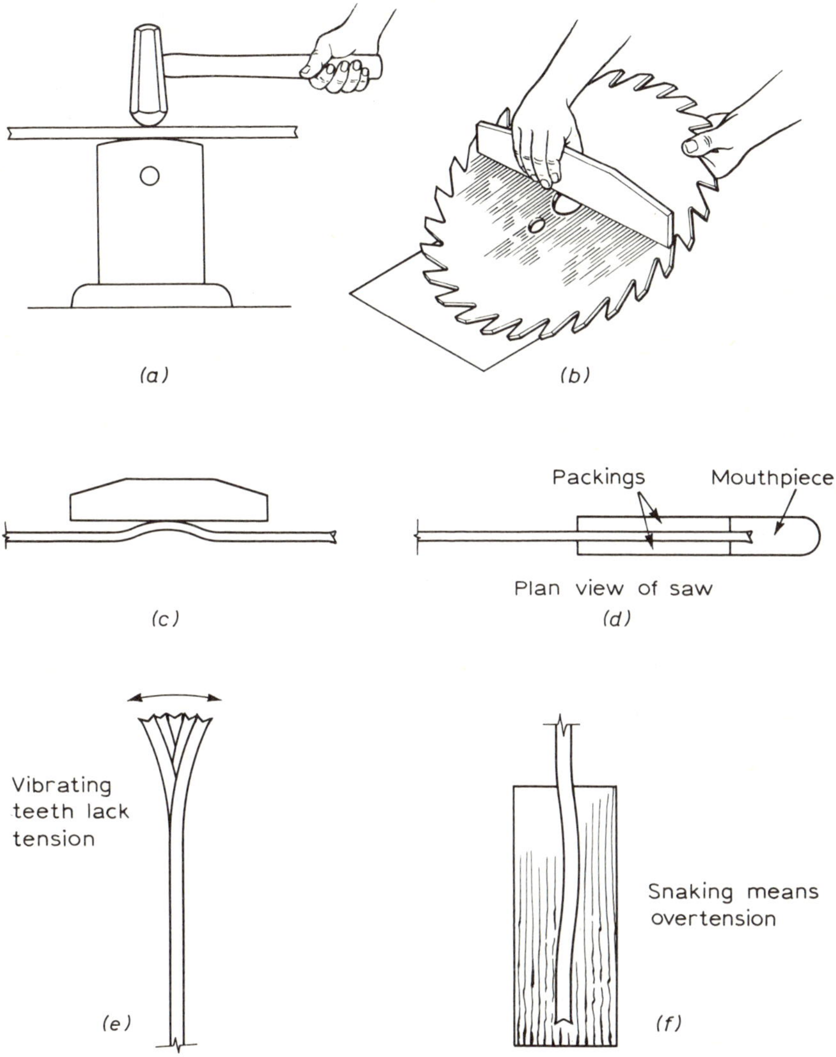

Figure 1.6 Tension and tensioning. (a) *Tensioning using dog-head hammer and crowned anvil,* (b) *tension in a saw which is seen as 'light' under a straight edge,* (c) *lumps in a saw plate are found by using a short straight edge,* (d) *packings help the regulation of tension and to lubricate the saws; the mouthpiece holds the packings in place,* (e) *a saw lacking tension will rattle at the tooth and* (f) *a saw with too much tension will tend to snake along the length of the board*

proportion than the almost neutral area between collar and rim. Not being able to stretch out and pull straight the rim area will corrugate and the saw teeth will flutter and vibrate from side to side. Application of light pressure to the almost neutral area between rim area and collar will cause a rise in temperature and in turn, expansion. The rim area will now be able to pull itself back to a truly flat plane again. If instead of causing a rise in temperature the saw is hammered from collar to rim prior to use, a compressive stress is set up between collar and rim and this is only relieved and all stresses balanced when the saw is run at the correct speed and temperature. If stresses are balanced, the saw will stay flat and stiff, and will also cut true, because all forces will be in equilibrium.

The presence of a compressive stress can be seen if the saw is supported horizontally at two diametrically opposite points of rim. The centre will fall away from a straight edge, and will also do so if the saw is turned over. This sag is the condition referred to as *tension*. An alternative way of expressing it is to speak of 'An excess of metal towards the centre of the saw which is pulled out when the rim expands due to cutting'. However it is expressed, this tension is put in by the use of a 'doghead hammer' and a 'crown-faced anvil'.

Because of the changing diameter of the saw as it is sharpened, and the variation in the rim speed due to this, the sag or tension may become too much or too little, and the saw will 'snake' or vibrate. Tooth vibration and snaking at the rim occurs if there is too little tension, and 'dishing' occurs if there is too much, because the compressive stress is not fully converted into a tensile stress by rim expansion. 'Dishing' is a condition where the saw centre bulges from side to side during the cut, thus forcing the saw rim from side to side as the cutting load varies. The saw kerf then snakes along the length of the board being ripsawn or else 'binds' tightly to one side and overheats.

Finally on tension, machine saws of above 450 mm diameter have a means built into the machine by which the sawyer can overcome slight idiosyncracies of tension. On either side of the forward half of the saw there are 'packing boxes'. These contain either felt or plaited-hemp packings, saturated with oil, which lubricate the saw. These packings enable the sawyer to control tension by varying their tightness of fit, and therefore to create a controlled rise in temperature at either the rim or the collar. Vibration at the teeth can be cured by packing more tightly at the centre, and dishing may be cured by packing more tightly under the tooth. If the saw is correctly packed it will overcome minor tension irregularities and run true.

Figure 1.6 illustrates the various features of tension and tensioning.

The machines used for sawing

RIPSAWING

The most common form of ripsawing machine is known as the *push bench* because there is no method of feed other than by manual push. The machine has a fabricated or cast iron base topped with a large flat table through which the saw projects. The gap or slot provided for the saw is filled in when the saw has been fitted by a 'finger plate'. This is a heavy cast iron plate having hand holes for extraction. The front half (towards the sawyer) of the table slot and finger plate each have a built-in packing box lined with hardwood. The rear halves are also lined with wood but have no packing box. The wood protects both saw and table should the saw suffer heavy deviation in use due to overheating.

The saw is clamped between collars on the end of the saw mandrel the thread of which is left-handed so that the saw nut tends to tighten up as the saw rotates. To remove the saw the spanner must be knocked or jerked in the direction of rotation. The collars are recessed to give a flat, firm grip. Between the mandrel (or 'spindle nose' as it is sometimes called) and the outer rim of the collar, there is a drive pin that is screwed into the back collar, and in use, projects through the saw pinhole into the outer collar.

On the table, to the inner side of the saw and parallel to it, is the work fence. This fence may be set by the sawyer slightly out of parallel to cause the wood

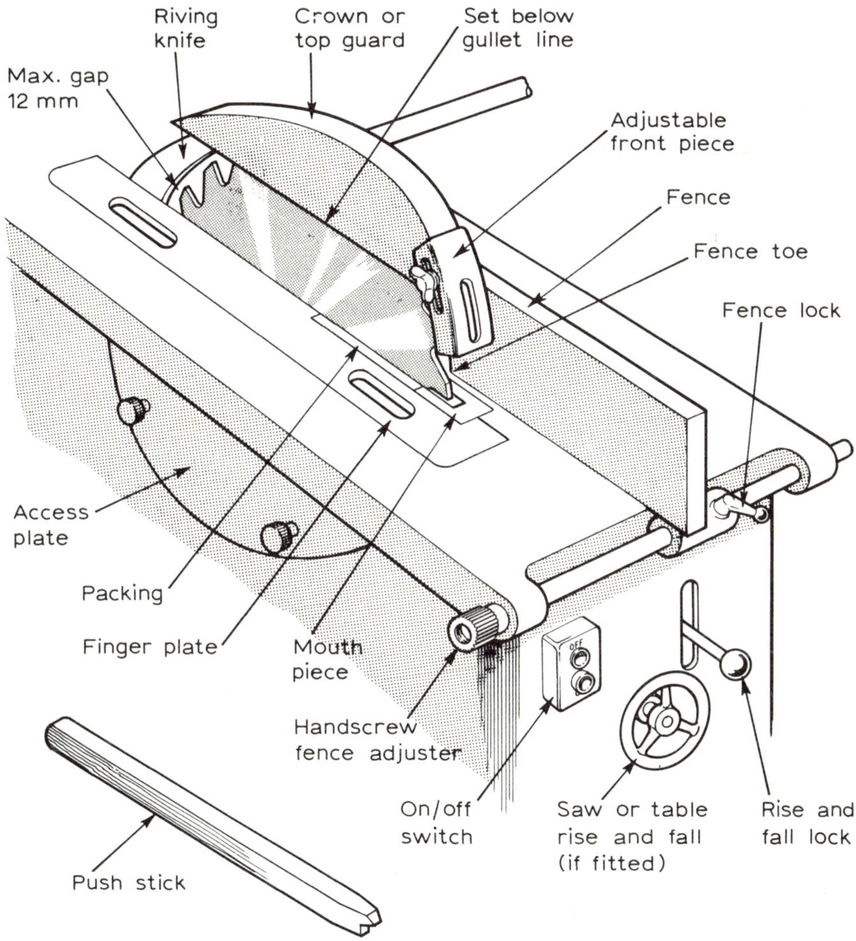

Figure 1.7 Circular push bench

being cut to hug closer to it; the fence and this situation is spoken of as 'giving a lead to the fence'. This lead will be approximately 0·5 mm for a saw of 660 mm diameter. The fence stands perpendicular to the work table for normal sawing. It may also be adjusted forward or backward to suit varying diameter saws. The fence front end or 'toe' is normally set at saw gullet level. If the toe is set too far along the saw, 'binding' occurs, i.e. the wood is pinched between saw and fence. The saw will overheat and run wild should this condition persist. If, on the other hand, the toe is not forward enough, the end of each board will have a 'jump' or projection, which will cause trouble when the time comes to plane or mould it.

The fence has a cross-slide to enable it to be positioned for the required width of cutting. Control for fine adjustment is by hand screw, after major positioning has been accomplished by free slide. Behind, and in line with the saw there is a 'riving knife', the function of which is to guard the saw back and to prevent the cut wood faces from closing in on the saw and being thrown back at the sawyer. The set position of this knife is subject to the Woodworking Machinery Regulations and so is the top or crown guard which covers the top of the saw and the outer side of the saw teeth. The crown guard has an outer flange extending downwards and a curved top cover for the top of the saw. At the front is an extension piece which has its own adjustment for height. The Woodworking Machinery Regulations are part of the Factories Acts, and are enforceable by law. Figure 1.7 shows, in diagrammatic form, the essential features of a circular rip bench or 'push' bench.

Heavier ripsaws have large powered rollers built in. These are known as *roller re-saws*. The rollers are weight loaded into the wood and rotate on the end of a fulcrum arm which allows the rollers to move in or out to suit variations in cutting width. The roller shaft may be pivoted to enable long boards to be fed tilted sideways should angular cutting be required such as for feather-edge boarding. A gearbox gives a feed range from 5 to 55 m/min. Any power-fed saw must have meticulous attention paid to the condition of the saw plate, the maintenance of the saw teeth, and to the setting of the fence and packings. The roller re-saw is found mostly in mills specialising in wood trim, fencing and packing-case material. They operate most economically on long softwood boards, but have now been virtually superseded by the band re-saw.

PRODUCTION MACHINES

For production sawing in shops other than sawmilling, the straight-line-edger or 'flyer' as it is sometimes known, is mostly used. This machine is so heavily guarded around the saw area that the requirements of the Woodworking Machinery Regulations are not usually enforced with regard to the riving knife. However, safety in design is covered in law by the requirements of the general regulations relating to the design of dangerous machinery. The 'law', when involved, uses as a basis the judgement of an imaginary 'reasonable man', and this is normally sufficient to ensure built-in safety.

The 'edger' has a very large work table and the saw may be mounted above or below. Along the centre of the table there are machined slides on which runs an articulated track, or, in the case of the under-mounted edger, two tracks, one to either side of the saw. The top surface of the links of these moving tracks are serrated and project approximately 3 mm above the table surface. Over-mounted saws run in a groove cut into the top surfaces of the moving track links. These grooves are dovetailed in section and hardwood fillets are knocked firmly into each link. In this way the saw is run with its lowest arc of periphery lower than the link top surfaces, and any board fed to the saw will be cut into two, yet the saw will not contact the metal of the link. Wood being cut is held down to the feed chain or track by pressure rollers mounted either side of the saw. On under-mounted machines, the track is split into two, and as the saw projects between them there is no need for the packing strips or grooves. On these under-mounted machines the top pressure rollers are all of the same size and are friction driven by the passage of the top surface of the wood being cut. Their purpose is to ensure a rigid grip between track upper surface and the underside of the wood. On over-mounted machines the pressure rollers are of two sizes. Small ones are set close to and at either side of the saw, and these are friction driven and four in number. Larger rollers are positioned both at the front of the machine and at the rear, and these are power driven. The working safeguards are two-fold. At the extreme front of the top pressure unit there are two sets of fine, steel fingers that are pivoted so that they always have their lowest extremities riding just clear of the track and pointing towards the direction of feed. The board being fed lifts as many of these fingers as is necessary for its width, whilst the outer ones remain hanging down. As soon as the tail end of the board passes through, all the fingers drop down, thus preventing any thin strips of wood, or 'off-cuts' being thrown back at high speed. These potentially dangerous and spear-like pieces are known as *fly-backs*.

The principle of edger design, which is indicated by the name of the machine, is that of straight-line cutting. The wood is pressed firmly down to the feeding track, and as this is guided by the machined slides, the wood is carried forward through the machine in a perfectly straight line. A fence is provided to enable set widths to be cut. With waney edge boards the sawyer lines up the board by eye to give a clear cut edge with minimum waste.

MULTIPLE SAWS

Essentially the same as the over-mounted edger is the multiple straight-line edger that may be set with a number of saws spaced by collars, and mounted on a long saw mandrel. Thus at one pass each board may be ripsawn into several pieces. The principal difference in design lies in the work track or chain, which is arranged to 'dip' below the saws to avoid contact.

CROSS-CUTTING SAWS

The most common of these is the straight-line pull-out type, with the saw mounted above the board to be cut, which lies on a roller table having a back fence and adjustable length stops. The operator assesses the board by eye and positions it against the length stop most appropriate to his cutting list and the board condition. The saw unit is then pulled by hand across the board, cutting it to length.

The saw is directly driven from a motor mounted at the front of the sliding machine head, and is mounted on an extension of the motor shaft. Variations in design allow some machines to pivot up to 45° to the right or left, but not both sides to this extent because as the saw is pivoted in one direction, the motor end moves in the other. The machine head may also swivel in plan to enable mitre cuts to be made. A wide range of cutting tools may be fitted that allow for trenching, notching, moulding or grooving. The higher output varieties of these machines have a power feed with foot control for the operator of a hydraulic circuit, but they loose their flexibility for angular work. Others may be fitted with automatic selection of work length stops to the extent that some commence their cutting strokes when the board end contacts its chosen stop. The operator of this type of mechanised complex sits at a control panel and controls board feed and length selection by push button.

THE DIMENSION SAW

A dimension sawing machine cuts wood to precise dimensions. The machine design allows for an almost completely universal form of cut. Sawing may be ripsawing, cross-cutting, angular sawing, compound angular sawing, panel sizing, grooving, rebating, tapering and as many variations in cutting form as the operator has the ingenuity for which to jig.

In general appearance the dimension saw is similar to a smaller push bench; it has the same rip fence and saw guards, but includes also a cross-cutting fence and rolling table. The saw may be raised and lowered, and may be tilted up to 45° towards the fence.

The rolling table consists of that part of the working table that lies to the outer side of the saw. This table is mounted on precision runners that allow the operator to push it forward or pull it back very easily. To this table is fitted as required the cross-cutting fence and stop bar, which may also be used for angular cutting. All standard angles are marked and the cross-cutting fence has a spring-loaded plunger. The rolling table may also be firmly locked for ripsawing work. To fit the saw the rolling table is moved away.

GENERAL

The best way for a student to familiarise himself with various machine designs, appearances and capacities, is to study as many manufacturers' catalogues as can be obtained. Tool catalogues give more information on range and design of small tools (fitted saws and cutter-blocks, etc.) than any textbook can hope to do.

Circular saw calculations

Saw calculations are mainly concerned with maintaining the optimum peripheral speed for each saw diameter. Ideally, this would mean having an infinitely variable adjustment to the revolutionary speed of the saw mandrel, which is an economic impossibility, as well as a potential hazard in unskilled hands. In practice, sawing machinery designers give a slightly high rim speed to new saws at maximum diameter, leaving to the saw doctor the job of stiffening-up the saws as they lose diameter. Referring back to 'tension', consideration of basic theory will show that 'stiffening' a saw will mean giving it less static tension (i.e. fall at centre) as its diameter increases for the same revolutions per minute. The sawyer may also regulate overtension by packing more tightly under the tooth.

The only situations in practice where saw calculations of this type become necessary are where it is felt that some irregularity in cutting is due to incorrect rim speed, or, due to the purchase of a second-hand machine, a motor or saw pulley has to be selected.

EXAMPLES OF CALCULATIONS RELATING TO CIRCULAR SAWS

In accordance with the Système International d'Unites (S.I. units) the unit of time is the second (s) or the minute (min) and the unit of length is the metre (m) or the millimetre (mm).

The correct speed of rotation R in rev/min for a given diameter circular saw is given by the expression

$$R = \frac{1000\,P}{\pi D}$$

where P is the required speed (m/min), D is the saw diameter (mm) and π (the Greek letter *Pi*) is taken as being 3·14.

The formula takes the required rim speed (metres per minute) and multiplies by 1000 to convert this to millimetres per minute. The saw circumference (πD) in millimetres is then divided into this to give the necessary number of revolutions per minute. Exactly the same formula is used if the rim speed is given in metres per second and the speed of revolution is required in revolutions per second.

Example 1.1

A saw of 720 mm diameter is required to run at a rim speed of 3000 m/min. Find the correct number of revolutions per minute to give this speed.

$$R = \frac{1000\,P}{\pi\,D} = \frac{1000 \times 3000}{3{\cdot}14 \times 720} = 1326{\cdot}95 \text{ rev/min (i.e. } 22{\cdot}1 \text{ rev/s).}$$

To find the rim speed produced by a given diameter saw at known revolutions the formula is transposed to give P in terms of D and R.

$$R = \frac{1000\,P}{\pi\,D} \quad \text{and} \quad P = \frac{\pi\,DR}{1000}$$

Example 1.2

A saw of 720 mm diameter is running at 1326 rev/min. Check by calculation that this is producing the correct cutting speed of 3000 m/min.

$$P = \frac{\pi DR}{1000} = \frac{3{\cdot}14 \times 720 \times 1326}{1000} = 2997{\cdot}82 \text{ m/min (i.e. } 49{\cdot}9 \text{ m/s).}$$

Allowing for rounding-off decimal figures to avoid unnecessarily laborious calculation, the revolutions given are correct, and this fact acts as proof of the formulae.

These two examples show that the problem of π is no less with metric calculations than it is with common fractions. The solution may lie in the use of logarithm tables, but this would depend on the ability and attainment of the individual student.

Gullet depth proportional to pitch is found by simple calculation involving the multiplication of tooth pitch by its ratio with the gullet depth. *The Handbook of Woodcutting* (H.M.S.O.) gives standard ratios that vary for each class of cutting.

Example 1.3

Find the necessary depth of gullet for a tooth pitch of 40·32 mm where gullet depth to pitch ratio is 0·375. Gullet depth D is given by the formula

$$D = RP$$

Where R is the ratio given and P is the pitch. Hence $D = RP = 0{\cdot}375 \times 40{\cdot}32 = 15{\cdot}12$ mm.

Power

Power is expressed in watts. Most sawing requires power to the extent of 746 W (1 h.p.) for 25 mm of cutting depth. This can only be arbitrary, for motor power is often related to the machine price and to the conditions under which it will be used.

2 Machine Planing

Cutting theory

All chopping is cutting; with wood, all cutting is also chopping, with blow after blow being struck and chip after chip being removed. A good hand plane used with skill produces a long shaving direct from the edge of the plane iron. (The expression *iron* is still used although cutting tools are mainly high-carbon steel.) The length of the shaving may well be the full length of the wood being planed. With machine planing, this cannot be—the planed surface is built-up cut by cut as the wood being planed is fed past a rotating cutter block carrying 2, 3, 4 or more cutters.

Each cutter strikes the wood in a downward direction (relative to the wood face), carries along around an arc and emerges pointing upwards. If the wood feed is stopped the cutter will leave a mark shaped to a perfect arc of the block cutting circle. The feed being constant, the cutter mark is no longer a true arc but an extended arc, known (in geometry) as a *curtate trochoid.* Generally the marks are assumed to be arcs. As each cutter succeeds its forerunner part of the previously cut arc is removed. The final surface consists of a number of closely spaced arcs, merging with each other to give the appearance of a truly flat surface. The separate arcs can only be seen with a good reflected light and even then the degree of visibility will depend on the combination of a number of variants. Figure 2.1*a* shows the enlarged view of a finished rotary-planed surface.

The factors that combine to make possible this semblance to a truly flat surface are:

1. The number of cutters on the block set to an identical cutting circle. (The degree of tolerance acceptable here is very low and would be measured in hundredths of a millimetre.)
2. The speed of revolution of the block.
3. The speed at which the wood is being fed past the block.
4. The design of the cutters and the block.
5. The design and setting of the machine chip-breaking aids.
6. The nature of the wood (species).

The first three of these factors are variable in magnitude and so lend themselves to use as factors in a simple form of calculation that enables planed surfaces to be graded according to the pitch of the cutter marks. The longer the pitch the less flat will be the surface. The pitch is measured from crest to crest of each arc. Planed surfaces were graded by the number of cutter marks to one inch under the Imperial system, but with S.I. it will be more reasonable to refer to pitch distance in millimetres.

The last three of the factors listed are variables which a skilful machine woodworker blends into the finished product according to his degree of skill.

CUTTING ACTION

The cutting action is, as said earlier, one of chopping. A chopper is a wedge-shaped knife which, when struck at the end of an upturned log, penetrates and divides it. The two cut faces of the wood slide up the chopper sides with an ever widening gap. The chopper edge no longer contacts the wood once it has entered. The wedge shape of the chopper rives apart the grain fibres until one piece parts from the other. Each cut face will be jagged and torn, the degree of this accords to the nature of the wood grain (factor 6).

A machine-planed surface cannot tolerate such erratic riving, and cutter and block design together with machine design (factors 4 and 5) combine to cleave the wood grain in such a manner as to leave an acceptable surface. The cutter is carried down along its circular path into the wood end as presented. It then travels on, swinging through horizontally and up again and out. As the tip commences to cut it splits away a chip. The chip is forced away from the parent board and up along the cutter face which is executing an ever increasing tearing effort on the grain fibre. As the cutter swings up to leave contact the chip has either the choice of being cut by the cutting edge or to rive deeply along the grain line and lift as a long, heavy splinter. Whatever happens, it must come away from the parent board. Figure 2.1*b* illustrates the cleaving effect of a rotating cutter.

Factors 4 and 5 determine the extent to which long-grain riving occurs. The limited cutter projection beyond the block face and the shape of the block face cause the partly severed chip to bend back on itself causing cracking across the width. This reduces the leverage it can exert on the parent board and makes long-grain riving less likely. The machine design helps here. Immediately in front of the up-going part of the block there is a heavy, metal chip breaker/pressure bar. This bar is set to exert its maximum pressure just in front of the line along which the cutter will emerge. The developing, long, heavy splinter is abruptly ruptured (after being weakened by the block chip-breaker design), by the close set edge of this machine's chip breaker, which tends to compress the grain fibre of the board to prevent splits from developing in the wood not yet in contact with the cutters.

Although the anology of a cutter as a chopper is in principle correct, in detail it fails, for although the cutting effect is caused by wedging and riving, the cutter edge is always in contact with the wood being planed throughout the length of each chip. This fact brings in yet another field for the application of skill. The cracking of the potential chips is accelerated by a reduction in the angle at which the cutter hits the wood. Block design is fixed, but by grinding an angle in reverse on the cutter an effective alteration is made in approach angle to the wood. The riving effect is less likely to occur because the chip is being forced back along the board rather than up and away from it. Also, the chip is turned back on itself more quickly thus causing it to crack more easily. Figure 2.1*c* shows how the combined block and machine chip breakers cause rapid severance of the chip, and Figure 2.1*d* shows how the secondary bevel on the cutter prevents a deeply riven grain.

BROKEN GRAIN

The broken grain effect often found when planing brittle wood is caused because the emerging chip has developed sufficient strength to rive itself away from its parent board way beyond the point of contact of the chip-breaker pressure bar.

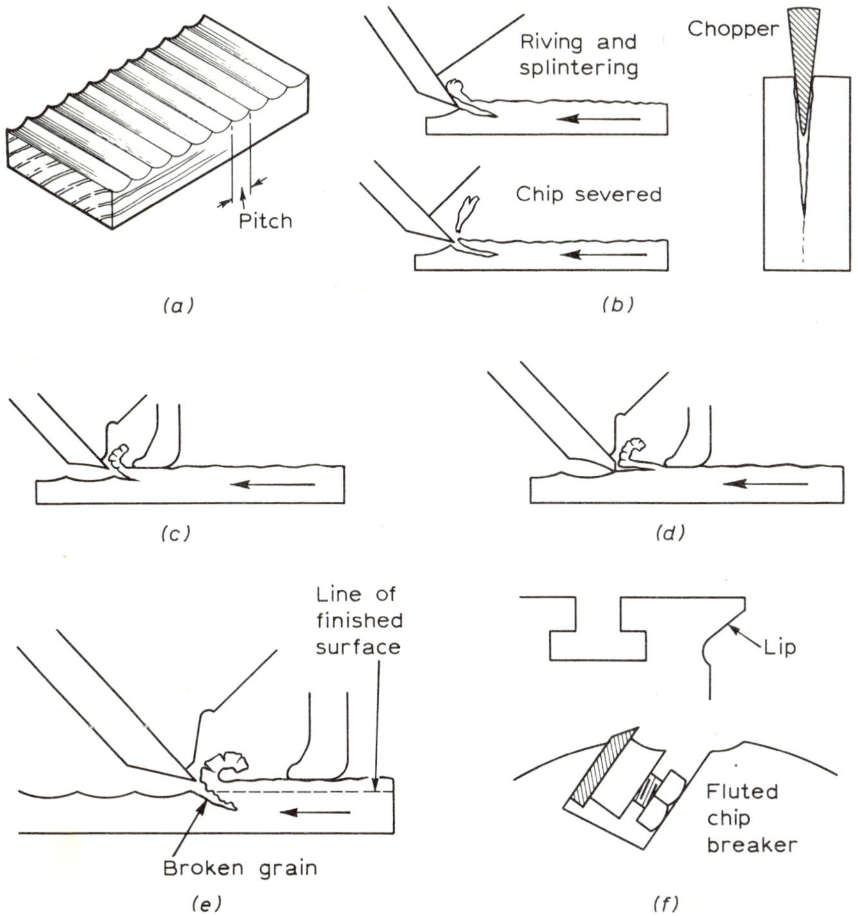

Figure 2.1 Principles of surface planing. (a) *Surface produced by rotating cutter block,* (b) *forward riving and chipping action compared,* (c) *lip on a square cutter block turns and cracks the chip and the machine chip breaker resists forward riving,* (d) *cutter with a face bevel turns chips more easily and reduces the depth of the forward rive,* (e) *if forward riving does occur to any extent, broken grain is shown on the finished surface and* (f) *square cutter blocks for planing have extended lips to turn chips, and circular blocks have a fluted form of chip breaker*

If the grain direction is hostile then the resulting tear or rive will extend deeper than the line of surface of the finished planed face. The second angle on the cutter assists to keep the tear or rive above the required line of finish. Figure 2.1*e* shows how the grain tear is built up.

The decision to use this second angle, known as the *face bevel* requires consideration of factor 6—the wood species. If face bevel is used, much greater power is required, which may lead to lower feed speed; if used for unsuitable timber (soft and woolly) the grain will be depressed and will react within a few days of planing and rise again. Cutters will also lose their sharp edge more quickly, because the chip, being kept short, will not be able to rise away from the cutter tip (as in chopping) and will cause much greater wear on the cutter edge.

Face bevels may only be used on power-fed machines, because the thicker edge and quicker chip rupture demands more power than hand feeding can apply.

Technical details of blocks

Hand feed machines must, by regulation, be fitted with a circular cutter block. Power planes may have circular or square blocks. Hand powered planers (surfacing planers) most frequently have two cutters, but may be found with either three or four. Power fed machines such as finger feed surfacers, thicknessers, panel planers, moulders and fast flooring machines, may have 2, 4, 6, 8, 10, 12 or even more cutters. The usual range is from four to eight cutters. When square blocks are used the cutters are limited to four. Square blocks are mostly lipped when designed for planing. This means that the face of the block immediately under the sharp edge of the cutter is built out to reach almost to the cutter edge. This lip is the chip breaker, turning and cracking the emerging chip to give a better surface finish. Circular cutter blocks have a flute-shaped groove along their full length immediately in front of the cutter tip. This serves the same purpose as the lip of a square block and both are shown in Figure 2.1*f*.

Cutters are held on square blocks by nuts screwed on threaded studs that are themselves screwed into the block face. The cutters are thick and have slots to fit to the studs. They also have a thin high-speed steel face welded or brazed to the bulk of the cutter which is of mild steel or carbon tool steel. The high-speed steel provides a long-life cutting edge.

Circular blocks have various methods of cutter holding. The most common is the wedge bar system, where the block has a dove-tailed groove machined for each cutter. Into this groove drops the thin h.s.s. (high-speed steel) cutter and a wedge bar. Along the length of this bar are several small set screws with square heads. Once the cutter and bar is in the slot the screws are unscrewed from the bar to the limit allowed by the walls of the groove. This pinches the cutter and clamps it. Variations on this have the screws threaded into the block body itself, and in this design, cutters are held when the screws are screwed in, pinching the wedge bar and cutter tight. Blocks having only two cutters may be 'slab' blocks, where two opposing cheeks are formed separate from the main block. These are bolted through to the main block, clamping the thin h.s.s. cutters in

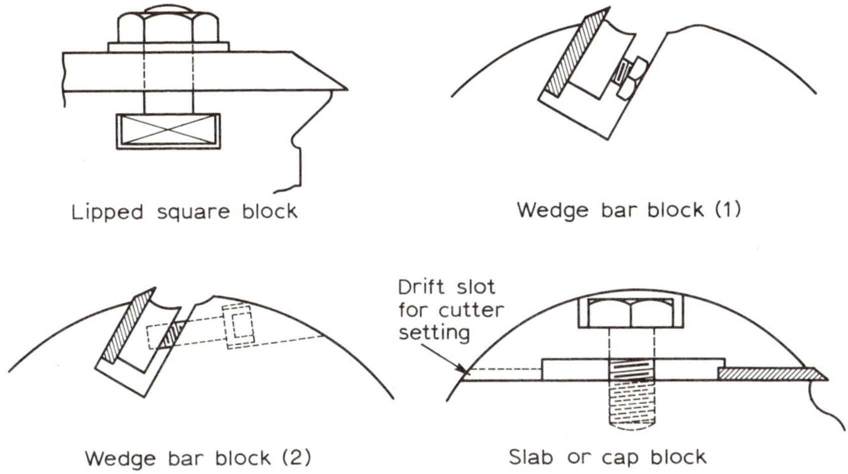

Figure 2.2 Cutter holding methods

the process. Figure 2.2 gives details of four types of cutter holding methods.

Setting adjustment to square blocks is by hammer tapping, but most modern circular blocks have some form of mechanically aided adjustment. These will be explained later. Another feature of modern machines is that many of them have the cutter set in the block at a slight angle to the block axis. This still keeps the cutter edge set in the line of the surface cylinder generated when the block rotates but does give a small amount of shear cutting. The value of this is theoretical and the aim is to cut brittle grain more easily.

CUTTER BLOCK ANGLES

The expression *cutter block angles* covers several angles pertaining to planing. These angles are:

1. The natural cutting angle.
2. The grinding angle.
3. The sharpness angle.
4. The clearance angle.
5. The face bevel angle (if any).

These are defined as follows:

1. The *natural cutting* angle is the angle contained by a line drawn along the cutter face, and a line drawn from the cutter tip to block centre. (This is determined by the angle at which the cutter seat is ground on the block and by the cutter projection.)
2. The *grinding* angle is the angle contained by the lines of the cutter face and the bevel of the ground portion.
3. The *sharpness* angle is that at which any sharpening is done in relation to the cutter face.
4. The *clearance* angle is that angle contained between the sharpness bevel (if

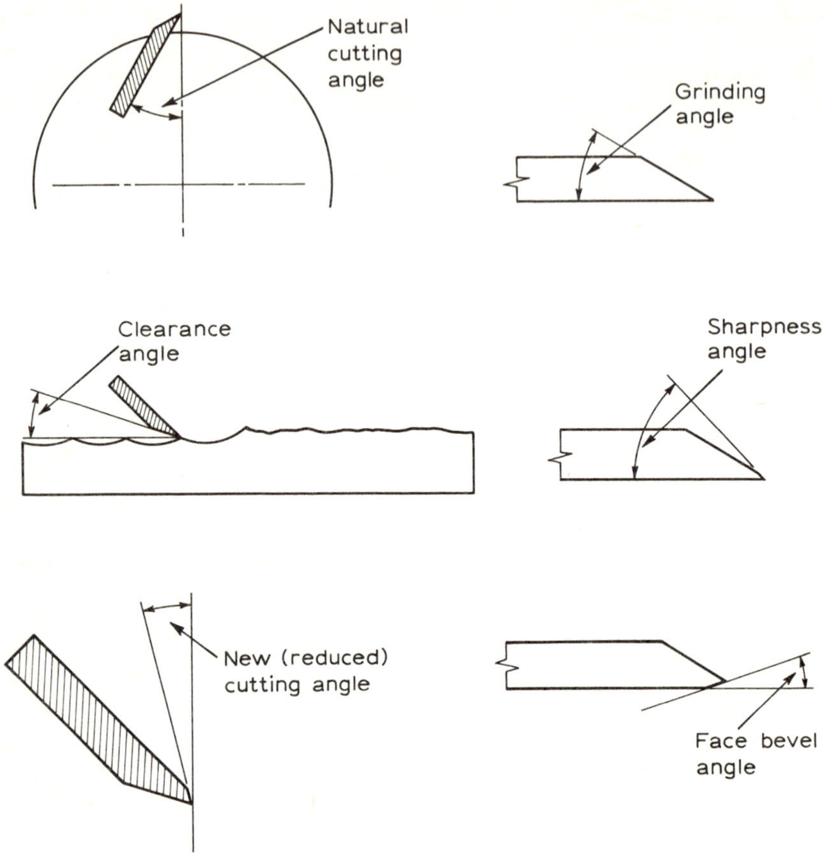

Figure 2.3 Various angles relating to planing blocks

any) and a tangent to the block cutting circle formed by the wood face.

5. The *face bevel* angle is that angle contained between a line drawn from the cutter tip along any bevel ground on the cutter face, and a line drawn along the cutter face.

Note: where a face bevel is used the natural cutting angle ceases to be the cutting angle, which will now be that angle contained between a line drawn along the face bevel, and a line drawn from the cutter tip to the block centre. This angle will always be less than the natural cutting angle and may even be zero degrees where the two lines coincide. Figure 2.3 illustrates these angles.

Summation of cutting theory

Surfaces planed by rotary cutters consist of a series of ridges formed by arcs of the cutting circle. The surface appears nearer to true flatness as the length (pitch) of the arcs is reduced. Variation in the number of cutters, feed speed and block revolutions will alter pitch length, and using three of these factors, pitch length or feed speed may be calculated.

The wood removed is removed by riving as each cutter acts as a wedge driven into the grain. The chip will tend to lengthen and the riven surface to be left rough, but this is controlled by the shape of the block face which acts as a chip breaker, and turns the chip back on itself with consequent weakening. Set in front of the block is the machine chip breaker that bolsters the board surface and holds the chip whilst it is severed by the cutter. With brittle wood even this arrangement will not prevent all chips from riving to a depth below the required finished surface. To avoid this the effective cutting angle of the block is reduced by grinding a face bevel on the cutter. This changes the direction of shear from down into the grain to along the board parallel to its face.

Principles of machine design

As with the hand process of bringing wood to size there are two stages in machine work. One face of the piece of wood must be made flat and clean, and one edge must be made straight, and square to the flat face. The next process will make the piece of wood parallel in thickness and in width, to the correct dimension.

SURFACE PLANING

The machine most used to bring wood flat and straight is the surface planer, sometimes known as the *overhand* planer. The essential basis of these machines is that there are two long, flat tables parallel to each other and placed end to end. Between the ends there is a circular cutter block. On the in-feed table there is a vertical fence set parallel to the table edge. This fence may be canted to any angle up to 45° relative to the table surface. This is to enable bevel edges to be machined.

The two work tables (in-feed and out-feed) may be raised or lowered on slides set tangentially to the block cutting circle. This gives the facility for control of cutting depth (in-feed table) and for correct setting (out-feed table). The tangential slides enable rise and fall to occur without the risk of the tables being wound into the cutter block (see Figure 2.4*a*).

The design principle is as follows: a piece of twisted wood placed on the in-feed table will find its own level, assisted by slight pressure from the machinist. Having attained one position it is held there and fed along the table, which, being flat, causes the wood to move in a straight line. At the inner end of the in-feed table the wood meets the cutter block which is set above the line of the table by the amount of cut required. The cutter block cuts away the high spots of the wood. These are the points on which it has rested. As feed continues (in a straight line) the out-feed table behind the block receives the planed surface and supports it. Because the out-feed table surface is always set level with the highest line of cut, the wood will continue to move in a straight path, guided by the two parallel, and flat, tables. All high spots are thus removed. Several passes may be required to produce a clean, flat face. The skill of the machinist will determine how many passes are necessary and how much wood may be removed and yet permit finishing to final dimension. The flat face is now held against the vertical fence and the lower edge of the wood is made straight and square to the face.

PLANING TO THICKNESS

Having achieved a flat face and square edge, hand woodworkers would use a gauge to scribe guide-lines indicating width and thickness. Machine planing avoids this. The machine used to make wood parallel and to a dimension is called a *thicknesser* or *panel planer*, with the very loose definition that thicknessers have square blocks and panel planers have circular ones.

The machine has a cutter block set above the line of the material to be planed. Horizontally below the block there is a rise and fall work table. To achieve satisfactory results the work to be planed must be held down on the thicknessing table. If the downward pressure is too great the degree of friction built up would inhibit feed movement. Set in the rise and fall table are two rollers with centre axes parallel to the axis of the cutter block and the surface of the rise and fall table. These rollers are mostly free running (friction-driven from the wood lower surface), but larger machines may have power-driven table rollers. This pair of rollers (called *bottom* or *anti-friction* rollers) is set slightly above table surface to break the frictional resistance between wood and table. The pair may be raised or lowered to suit the condition of the wood being planed, which may be hard, soft, rough or smooth. Arranged immediately above and parallel to these bottom rollers are two top rollers. These are power driven, and grip and feed wood through the machine.

When wood is fed into the thicknesser its top surface is rough and its lower surface pre-planed smooth on the surface planer. The feed roller in front of the cutter block has a serrated or fluted surface which enables it to grip the wood and feed it in.

To accommodate variations in board thickness throughout its length the in-feed roller is weight (the dead weight of the rollers) and spring loaded so that it may at all times exert sufficient pressure to cause its serrations to bite, and yet not indent the raised bottom rollers into the wood lower surface. Should such indentation occur feed will be irregular. In addition to this overall spring loading, large machines have the feed roller cut into sections, each approximately 25 mm long, and each internally sprung on the main roller shaft. Each section may rise to suit minor variations in wood thickness at the same time as the whole roller moves to suit the general configuration of the board being planed.

The cutter block would tend to lift the wood as it is met and tend to break its surface. In between the cutter block and the in-feed roller there is the chip breaker/pressure bar. This holds down the wood and provides an edge on which the chips can break. This pressure bar is also spring and weight loaded and may also be in sections of up to 75 mm in length, each individually sprung. If so, it is known as a *sectional* chip breaker.

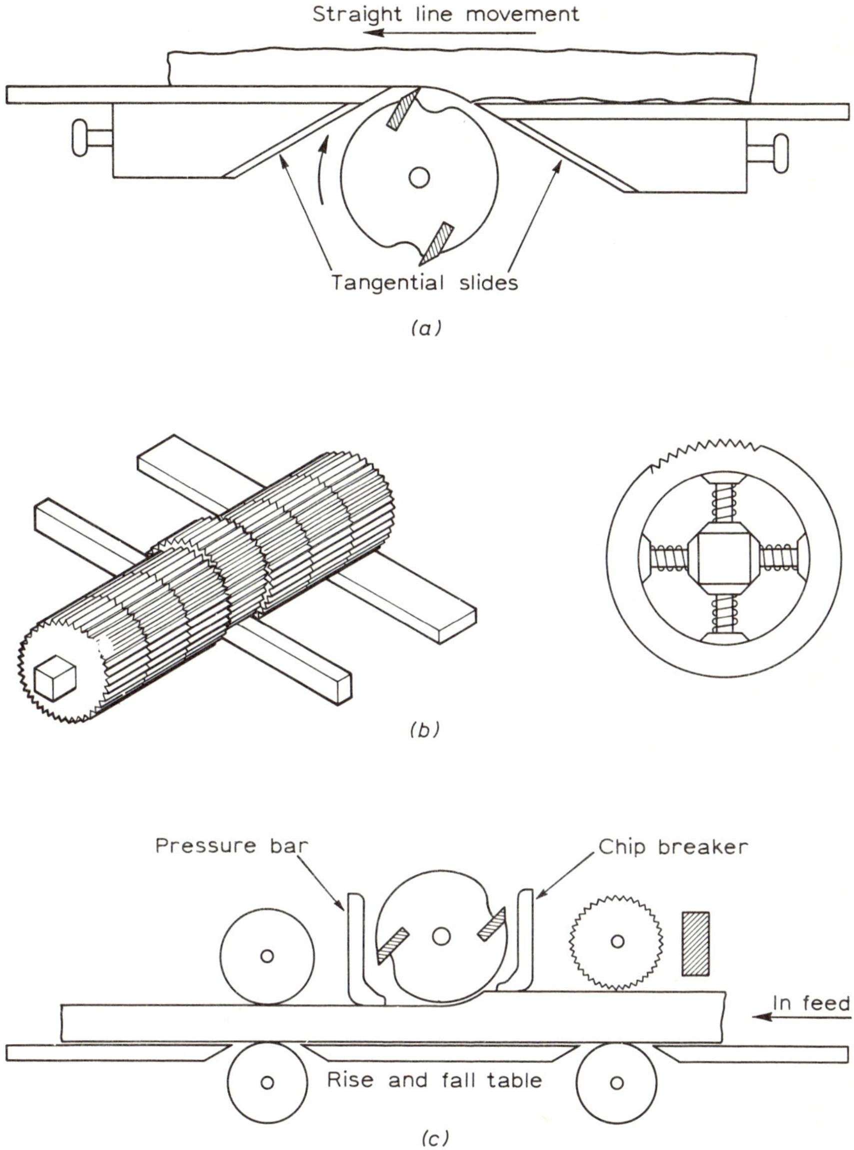

Figure 2.4 Mechanical principles of surface planers, thicknessers and combination machines. (a) *Straight line principle of the surface planer with tangential arrangement of table slides,* (b) *sectional feed rollers permit varying thickness pieces to be planed together; each section has its own internal springs, and* (c) *combination planer (which has the same design as the thicknessing planer) without the top surfacing tables*

A second top roller or out-feed roller, having a smooth surface, is set behind the cutter block. It is again spring and weight loaded, but not sectional. This roller pulls out the tail end of the board as it leaves the cutter block.

As each board projects out from the rear of the machine its over-hung weight tends to see-saw it on the table edge and back bottom roller. The tail end tends to rise and suffer injury by the cutter block. To prevent this another pressure bar is set between cutter block and out-feed top roller. This holds down the board tail end. This pressure bar is again spring and weight loaded. Feed speeds are varied to suit conditions of cut (light or heavy) in soft or hard wood. For this purpose a crash gearbox is usually fitted, which may or may not have a clutch to disengage feed. The feed unit may have its own motor, or it may be linked by belting to the cutter block motor. Figure 2.4*b–c* gives details of surfacer design, sectional feed rollers and feed arrangement of thicknessers.

COMBINATION PLANERS

Where feed and/or quantity of work does not allow the full use of two machines, i.e. the surfacer and the thicknesser, they may be combined into one machine.

On top are the two surfacing tables, with fence, bridge guard and cutter block, whilst below there are the rise and fall thicknessing table, in-feed and out-feed rollers and the chip breaker/pressure bar and rear pressure bar. To give chip clearance when thicknessing, the top tables may be pulled away from the cutter block. On no account must surfacing be performed with the table apart.

Due to the limited space available, the feed system is not complex as it sometimes is in a single machine thicknesser. Rollers and chip breakers are solid, not sectional, and tend to be smaller in size. Adjustment is rather more difficult for reasons of restricted access. Nevertheless, combination planers play a large part in the general wood-using industry, and produce accurate work economically.

Cutter and machine maintenance

Maintenance required to achieve quality planing involves cutter grinding and setting, roller and pressure bar setting, and the prevention of resin and chip build-up on table and rollers. Non-special maintenance will cover bearing wear, feed complex adjustments, and attention to the slides and gearing of the rise and fall table and the surfacing table.

CUTTER GRINDING

Whether cutters are ground within the factory or sent out to a grinding agency, they are most often ground in a wet-trough grinder. This consists of a large trough filled to a marked level with a mixture of water and soluble oil. The water is a coolant to prevent frictional heat developing and to disperse it should it occur. The oil prevents rust in the machine parts and provides the little degree of cutting lubrication that is necessary. Strictly, grinding, which is an abrading process, is directly opposed to the purpose of lubrication, which is the prevention of abrasion. Lubrication in grinding prevents undue friction between the bond material of the wheel and the material being ground.

Set in the trough is a pivoting cutter block, sometimes called the *dummy* block. Thin 3·2–4 mm high-speed steel cutters are clamped to this block, which is then rotated to the correct angle for grinding. The machine is indexed for this. High-speed steel (h.s.s.) is a carbon steel alloyed with tungsten, nickel and chromium in varying proportions. This gives a hard-wearing long-life cutting edge.

The usual grinding angle varies between 30° and 35°, with the latter giving a stronger edge for harder woods. The trough grinder will grind thick cutters for square blocks and it is sometimes an advantage (where frequent hand sharpening is shop practice between cutter changes) to have a hollow-ground grinding bevel. The trough grinder has facility for this. Along the top edge of the trough are two slides along which rolls a carriage carrying the grinding head and motor, and the traversing gear and motor. When grinding, the carriage is driven from end to end of the trough (limited by pre-set check switches to the individual cutter length being ground). At the end of each traverse, the traversing motor reverses and returns the carriage to the other end of its stroke.

The wheel is a resin bonded (waterproof) and wired (for safety) cup wheel, with aluminium oxide abrasive of grit size 36 to 48. Grit size and bond characteristics are chosen by the operator to give best results from his own method of work. Some men prefer a coarse wheel cutting quickly and leaving an edge to the slipstoned, whilst others prefer a finer grit wheel that takes more time but that produces a finished sharp edge.

The spindle on which the wheel is mounted is an extension of the motor shaft and is set, subject to adjustment, vertically in the machine. The wheel cutting face is downward. The nut holding the wheel may have an extended projection in the form of a large screw. This ploughs through the coolant and throws up a splash wave that washes the point of contact between wheel and cutter. An alternative design has a water pump that feeds coolant directly fore and aft of the stone/cutter point of contact.

As the grinding wheel is traversed it is automatically lowered by a pre-set amount. This may be set at 0·005 mm, 0·01 mm, 0·015 mm and 0·02 mm (0·0002 in, 0·0004 in, 0·0006 in and 0·0008 in). A total amount to be ground away may be selected up to approximately 0·25 mm (0·01 in). When the machine has ground this amount away after pre-setting at first contact, the fall of the wheel is arrested but traverse continues, giving a highly polished ground face. Overheating during grinding may produce microcracks in the cutter edge which can run into gaps when the cutter is used, or the cutter may bow due to expansion, be ground straight by the grinder, and take up a hollow shape when cold.

CUTTER SETTING

Good setting practice is aimed at setting all the cutting edges to the same cutting circle throughout their full length. Engineering diagnosis of the cutting situation holds that the achievement of such a degree of setting accuracy is beyond normal capacity. Such factors as bearing wear, slight unbalance of block and/or cutter, together with the limitation of accuracy of mechanical cutter setting devices, will defeat even the greatest care and skill.

This does not mean that great care need not be taken; the nearer the setting aim is to full achievement the better will be the cut, and this applies to all planing with rotory cutter blocks. A block set with four cutters all registering on the finished work (each cutter leaving its own arc mark) is possible—for some part of each cutter length. Most often, after very careful setting, each cutter will leave a mark on the wood surface, but of varying pitch. This degree of accuracy will leave a good surface, much better than one where one, or more, of the cutters are so badly set as not to cut at all. The finished surface will always show the mark of the cutter having the greatest projection and this will have the greatest arc of pitch. Cutters set very close behind this leading cutter will leave a smaller pitch mark. Cutters set very slightly back again will cut, but have their marks removed by following cutters. Finally, a cutter very poorly set will only cut air. Observation of work produced and of the used cutters will enable the setting situation to be diagnosed. A slipstone may be used to sharpen away the highest cutter in order to bring more cutters into the cutting orbit. The degree of tolerance in accuracy acceptable will vary with the feed speed. (Block revolutions and the number of cutters will also affect the surface, but these are outside the control of the machinist; feed speed may be selected). It has become accepted that under average cutting conditions, a difference in projection between two cutters of 0·5 mm will mean that the finished surface will carry only the marks produced by the cutter of greater projection.

SETTING METHODS

Surface Planer

Modern practice is to use mechanical aids for setting. Most common of these is a system where compression springs are set into blocks behind the cutters. The cutters are pushed into place, thus compressing the springs and are held by moderate tightening of the screws. A bridge device, registering from the bearings, the out feed table, or most commonly, the block surface, is placed over the block. On the underside (nearest to the block) there are two registration pads, one for either end of the cutter. When the holding bolts are released, the cutters are pushed out by the springs until their front edges are arrested by the registration pads. Cutter holding bolts are then fully locked. As each cutter registers on the same pad a fairly high degree of accuracy can usually be achieved. Wear and tear, dirt and moisture can all contribute to poor setting using mechanical aids. One form of the above device holds itself to the block by inserted magnets.

An alternative form of mechanically assisted cutter setting requires a device known as a *precision cutter setter*. This has a rise and fall pad which is spring loaded and which operates a lever registering against ruled calibrations. Each calibration is approximately 1·6 mm and the ratio of movement between pad and indexing lever is approximately 10:1. Thus a full movement of the index lever across one division shows that the cutter has raised the pad 0·16 mm. The

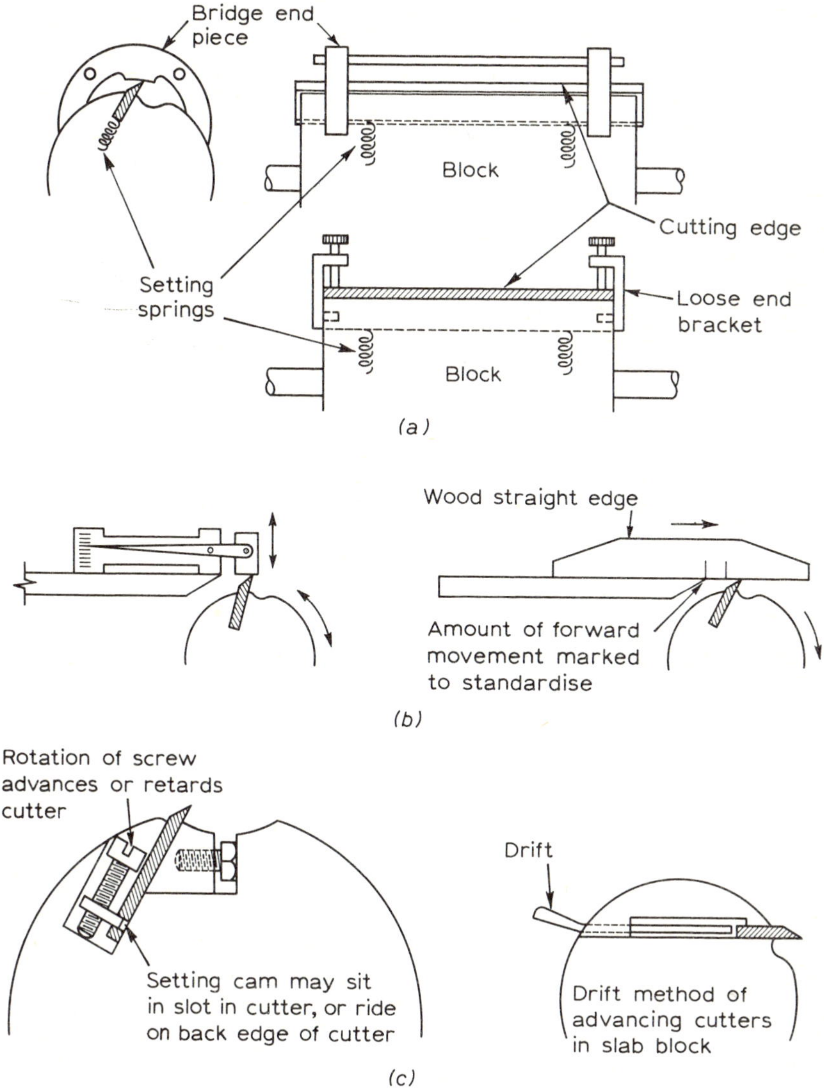

Figure 2.5 Cutter setting methods. Two methods are shown in (a) *using attachments for the limitation of projection where cutters are advanced by springs set in the block,* (b) *either a precision cutter setter or a wooden straight edge can be used to register cutters from the back table and* (c) *cutter block with built-in adjustment screws for advancing cutters to the required position. Blocks without these or without springs may have slots into which drift punches are inserted*

block is rotated by hand to cause movement of the pad and lever. Each cutter in turn is checked and set, at both ends and in the centre. Blocks using this method of setting either have set screws registering in slots at the back of the cutter for adjustment, or they have slots cut into the block through which a drift is passed for the purpose of tapping the cutter out. To reduce the projection of a cutter, it is knocked back from its front edge either by a steel punch on the heel of the grind or by a piece of hardwood on the cutting edge. Careful usage can achieve very accurate setting by this method, but the cutter centre must be set first because there are not usually more than two adjusting screws or slots, one at either end. A variation of this is to set cutters from the back, or out-feed table, using a wooden straight edge. Cutters are set and adjusted so that throughout their length they pull the wooden straight edge forward by the same amount when the block is rotated forward by hand. Figure 2.5 illustrates these various setting methods for the surface planers.

With all of these methods the aim is to set the cutting edges level with the surface of the out-feed table. If the cutters are set above the table line the work will tend to be round and have a 'jump' or 'dip' at its end (Figure 2.6*a*). If setting is too low, then either the work will not feed when it meets the table edge, or it will rise very slightly and thus produce a hollow edge (Figure 2.6*b*).

Thicknessers

Modern machines incorporate setting devices and the same limitations of accuracy apply. Wear and tear, dirt and misuse can defeat the machine designer in his efforts to achieve a method of simple, accurate setting.

With thicknessers and panel planers the devices are usually built in, leaving only the job of adjustment of cutters to pre-set limits to the machine operator. With mechanically fed machines the various pressure bars and rollers are set in close relationship with the block cutting circle. The rise and fall thicknessing table carries a dimension indexing finger. If the cutter-block cutting circle is varied this alters the relationship throughout the machine, and rollers, pressures and index finger will be incorrectly set. With the surface planer, block cutting circle is not of great importance because the first setting job is to set the out-feed table level with the highest point of the cutting circle.

The devices on thicknessers and panel planers are based either on small, touch-setting rollers, that are advanced to the cutting circle, or on contact pads or rollers carried on a bridge swung into position along the length of the block. The cutters, when fitted to the block, are advanced or retarded for setting by a similar means to the cutters on the surface planer.

If the machine has no built-in setting device then cutters may be set with the aid of a precision cutter setter, working from the rise and fall table below the cutters, and with the sensing pad uppermost. This is rather awkward because of confined space. Another method (and more popular) uses two pre-planed and parallel wood strips, approximately 750 mm × 100 mm × 32 mm.

The thicknessing table is lowered and the two strips are positioned through the machine, one standing at one end of the bottom rollers and the other at the opposite end. This means that both of their top edges are within the plane

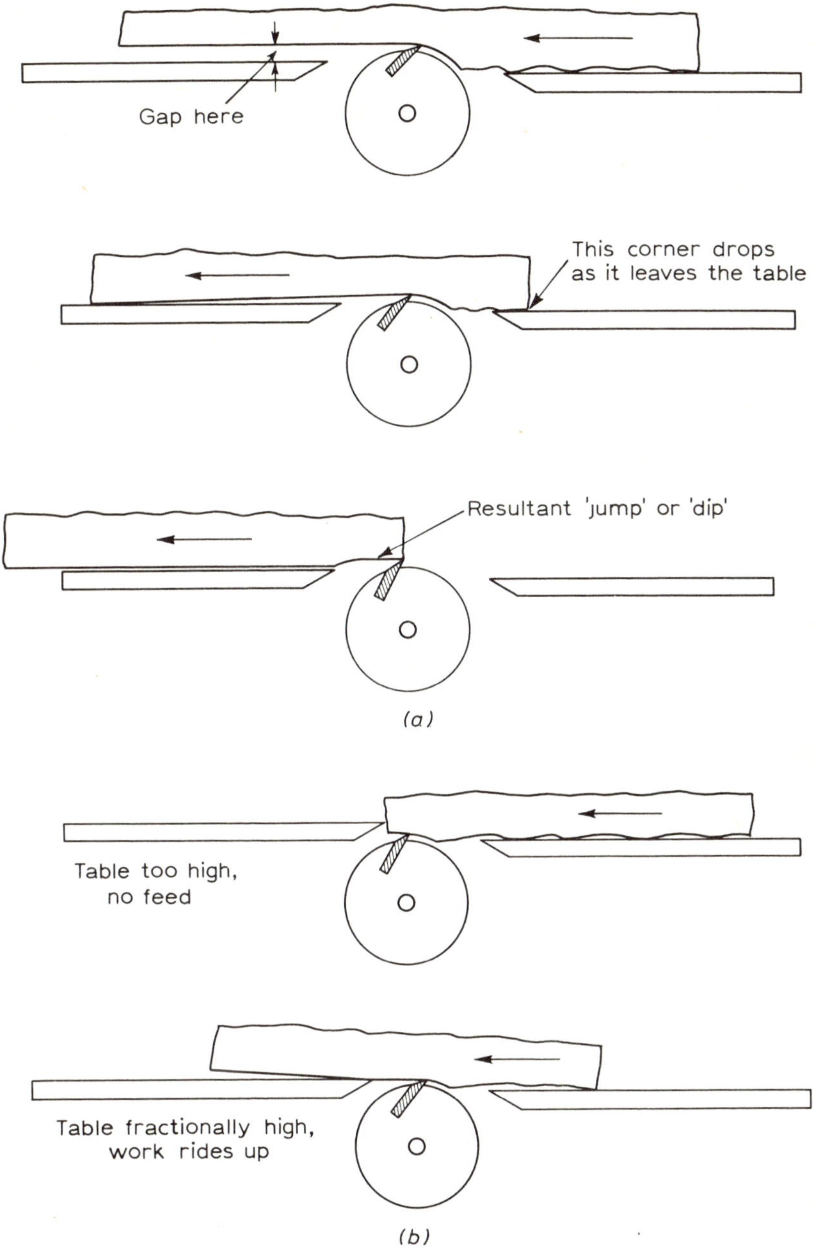

Figure 2.6 Faults in surface planer setting. (a) *Back table of surface is set lower than the line of the cutters and the tail end of the board being planed will drop and close the gap between it and the back table, and have a 'dip' or 'jump' mark left on its end.* (b) *Back table will prevent feeding if it is set much higher than the line of the cutter. If the error in setting is very small then the workpiece will rise and after being cut along the centre portion it will close down at its front end and will not be planed at its tail end*

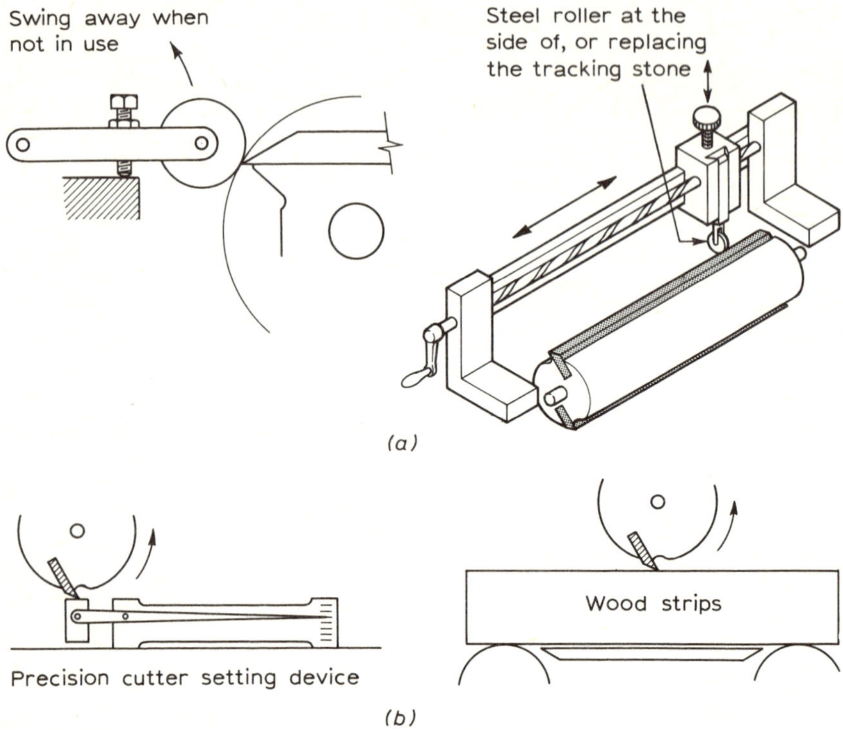

Figure 2.7 Cutter setting on the thicknesser. (a) *Two built-in devices to aid cutter setting. The tracking bridge may carry a setting roller, or this may be substituted for the abrasive stick.* (b) *The precision setting device can be used, but an almost universal method where built-in devices are not present, is the use of two parallel-planed wood strips with a light dusting of chalk on the top edge*

surface produced on any board planed on the machine. The cutters are then set just to brush the two top edges of the strips. A dusting of chalk on the strips' top edges makes it easy to see if the cutters are correctly set. No chalk on the cutting edge means no cut. Wood shaving on the cutting edge means too much cut. Setting is just correct when the cutters each pick up a dusting of chalk as the block is hand rotated.

When the method of setting for the cutters uses registration from the rise and fall thicknessing table, pre-setting of the machine bottom rollers is essential. These rollers carry the workpiece and if they are set out of parallel (or unevenly covered with gum and chips) cutter setting will be inaccurate. Setting methods are shown in Figure 2.7.

Machine setting

THICKNESSER AND PANEL PLANER

The setting requirements of these machines are the positioning of the feed rollers and bottom rollers, and the tensioning of their control springs.

The general method (which may be varied slightly according to the range of adjustments built in by the machine maker) is to use two wood strips as described earlier for setting cutters. The thicknesser table is lowered and the two strips are passed through the machine each resting at either ends of the bottom rollers. (Reference to Figure 2.4*c*, which gives a cross section through a combination planer will assist in the understanding of the method.) Below the bearing housings of the bottom rollers there are adjusting set screws. These are used to adjust the 'four corners' (for want of a better way of expressing it) of the bottom rollers until they are all evenly above the table surface by about 0·4 mm, although running performances must control final height. When set correctly, the bottom rollers will both move if either of the wood strips are pushed downwards and forwards. This will indicate that the rollers are doing their job of breaking the frictional grip between wood and table. When the rollers are set and locked, the table is wound up until pre-set cutters just brush the wood top edges when the block is rotated by hand. If the cutter setting is accurate and the bottom rollers are set true, all cutters will just brush the wood strips.

A satisfactory setting has the lowest edges of the in-feed roller (strictly not an edge, but more easily visualised as one) and the chip breaker/pressure bar set approximately 1·5 mm below the line of the cutting circle, and the out-feed roller and pressure bar set 1·00 mm below. The purpose of having roller and pressure lines below the cutting circle is to ensure continuous feeding should the board being planed run thinner in parts. This setting will also avoid the risk of thin boards striking cutters without being held by top pressure. The out-feed roller must of course be lower for it to grip at all.

Having set the table position to the point of cutter contact with the wood strips, the table is then lowered 1·5 mm. The four set screws, housed into the main casting at either side, are released and the in-feed roller and pressure bar are allowed to fall until they are resting on the tops of the wood strips. Set screws are unwound a fraction more and then wound up until first contact is made with the undersides of the roller- and pressure-bar support points. In this position they are locked with their lock-nuts. The table is then raised 0·5 mm and the sequence is repeated for the out-feed roller and pressure bar. The compression springs that apply downward pressure in addition to the dead weight of the rollers and pressure bars, must now be set. These springs stabilise momentum and prevent sudden upward jumps as the leading edge of the workpiece first strikes and raises each pressure bar and roller. The aim of the spring adjustment is to give sufficient pressure to ensure good feeding and control of the wood, yet not enough to compress it if narrow pieces are being planed on edge.

The method of setting is to unscrew all lock nuts and the spring compressing nuts until all springs are loose. The nuts are then screwed up until they first start to compress the springs, and from this point they are then screwed up another four to six turns. This should give sufficient compression, but as has been stated before, final adjustment may have to be made after running tests. All lock nuts must now be screwed up and locked firmly as they are subject to much vibration in use. Some modern machines have interlocked rise and fall in-feed rollers and pressure bars whereby the feed roller raises the pressure bar if the wood to be planed is exceptionally thick. With this system a thin steel gauge is used to determine the difference in setting and the point at which lift of bar occurs. It is useful to have access to the machine maker's instruction manual for roller and pressure setting routines, for apart from the general principles as outlined, there may be special techniques to suit individual machine design.

As soon as rollers and pressures have been set, a feeding test is made. Wood should feed through smoothly without irregularity. A piece once planed should feed through again at the same setting without roller indentation (but not soft species). If any irregularity should occur, observation will determine at which point it occurs and indicate the remedy. Faults are usually due to setting heights of bottom rollers and back-pressure bar.

Starting with the point of in-feed, the wood meets the first top and first bottom roller. If the top roller is too low or has too much pressure, it will not rise to the wood. If it grips but does not feed, the bottom roller may not be high enough. If the wood slews to one side, either the top or bottom roller may be out of parallel in height with the cutting circle.

Feeding through, the next point of stoppage may be the chip breaker/pressure bar. Again, if feed stops here, line of the pressure bar may be too low. However, it could be that there is not sufficient pressure on the in-feed roller, or the bottom roller is above the table surface but insufficiently so for the timber being planed. There may be too much compression on pressure bar springs. The wood will now meet the upcutting part of the cutter block. If it chatters, or, on later examination, has noticeable ripple marks, pressure on the chip breaker is most likely insufficient, or the bottom roller is too high and is producing a see-saw effect.

At the next stage, the only faults observable relating to the setting of the back pressure bar are the faults of too low a setting line, when wood will not pass beneath, or, too much pressure, when wood will pass, but then stick.

Out-feed top and bottom rollers may cause poor feeding if the top one is too low, or has too much pressure. In this situation, work will not feed on. If the top roller is too high, or has too little pressure the tail end of the wood will not feed out. The bottom roller (out-feed) will see-saw wood if set too high, and cause bad ripples on the end of the board, or cause the board to slew sideways if set out of parallel. End of board chatter may be due to insufficient spring pressure on the rear pressure bar.

Finally, reasonable feed that produces a surface that is apparently flat, can still show faulty setting errors if held obliquely to the light.

If both bottom rollers are set too high the planed surface will show a regular pattern on each board. Approximately 75 mm from the front end there will be a slight increase in board thickness. This will continue for about 100 mm. The board will then thin out again until the same pattern is repeated at the tail end. The regular steps are caused by the lift of the board as it rises to the bottom rollers and falls off. Figure 2.8 shows the cause and effect for several common feeding faults.

Jointing

TRACKING AND GRINDING IN SITU

Larger planing machines, having more than two cutters, may be fitted with a device known as the *jointer* or *tracker*. The origin of the name 'jointing' is

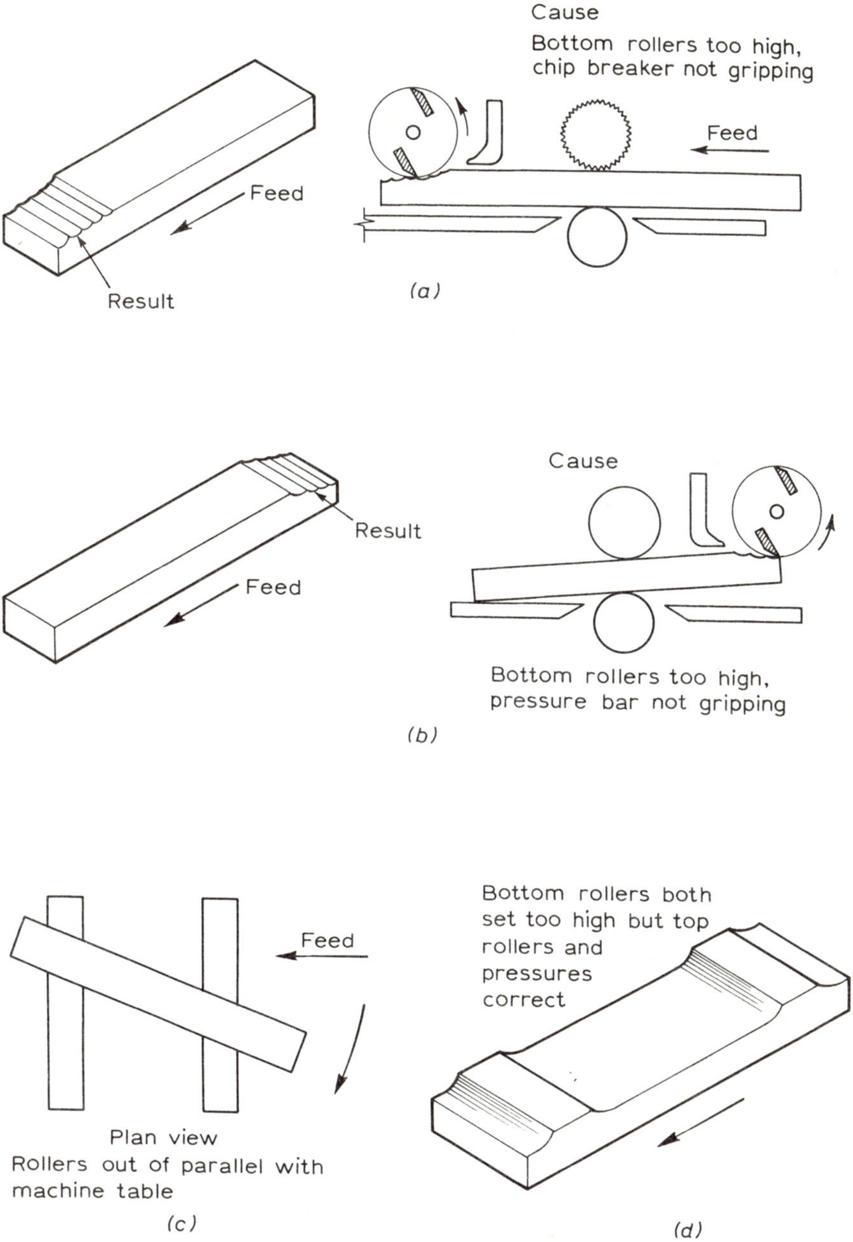

Figure 2.8 Roller and pressure bar setting faults. (a) *See-sawing on the first bottom roller occurs if the roller is set too high and the chip breaker is also too high or has insufficient spring pressure,* (b) *the same effect may occur if the out-feed bottom roller is too high and there is insufficient top pressure,* (c) *bottom rollers set out of parallel with one end lower than the table will cause the work to slew to one side and* (d) *if the regular pattern marking shown occurs, then both bottom rollers are set too high and must be lowered*

obscure, but the second name 'tracking' adequately conveys the meaning of the two terms. It is a process of applying an abrasive stone to the rotating cutter block for the purpose of reducing all cutting edges to a common cutting circle so that each will register on the finished surface. More important than this, however, is the fact that tracking gives the facility of resharpening the extreme cutting edges without the time wasted on resetting.

A cutter block having all cutters cutting and leaving their marks on the finished surface will cut more easily (smaller chips), leave a smoother surface (smaller pitch across cutter marks) and tend to cut for longer periods between sharpening (less work for each cutter). Additionally, as a marginal advantage, running balance may be improved. Theoretically, tracking should be done when wood is being cut as this takes up slack in bearings, but this is not practical.

Cutters are set carefully in the block and then, with the block running at full speed, an abrasive stone is fed along the line of the block and gradually adjusted inwards, until, with the block stopped, inspection shows a thin silver highlight on each cutting edge. The condition now is that within the limits of accuracy of the machine bearings, all cutters are rotating in the same cutting circle and should leave their mark on the finished wood surface, and feed speeds may possibly be increased.

Safety

Planing machines can be used safely, but danger is inherent. Accidents can be avoided if the operator takes care and the Woodworking Machinery Regulations are obeyed. The greatest danger arises from the use of the surface planer, especially when jobs other than straightforward planing are involved. Rebating from the table edge, chamfering and tapering are all jobs that call for great exposure of the cutters. Sensible application and use of guards, push sticks, and pushing blocks, together with back stops for dropping-on jobs, will make the job safe. When surfacing, the greatest danger lies in failure to use the bridge guard correctly. Wood being planed should be long enough to be held with safety, and always be passed below the bridge guard and between the end of the guard and the fence for edging.

Calculations

Surface finish has traditionally been graded by means of the number of cutter marks per inch showing on the surface. Too much importance should not be attached to this measure however, for quality hardwood joinery and furniture materials are invariably sanded or hand finished. A good surface is immediately recognisable, and so is a poor surface.

The grade code used varies, but mostly surfaces having eight or less cutter marks per inch are regarded as poor; finishes of 8–12 marks are acceptable for non-obvious joinery, and that which is to be external and painted. Interior painted work calls for as many as 18 cutter marks, whilst hardwood joinery and furniture that is to be cleaned and polished is most economically handled for further processing if it starts off with as many marks as 24.

With the adoption of S.I. units, new standard forms must emerge. It seems reasonable to work to pitch distances rather than to select an arbitrary unit of length and relate the number of cutter marks to this, but for a while it will be sufficient to think in terms of 'cutter marks per 25 mm'. Remembering that cutters may be set closely to a common cutting circle and yet not leave a mark on the finished surface, it would also appear reasonable to work any related calculation on the basis of one cutter per block. There will be other cutters and they may all cut, but under certain conditions, as defined earlier in the section on the theory of cutting, only one cutter will record its mark on the finished surface; others will have their mark removed by the cutter having the marginally greatest projection. This situation is known as *one knife finishing*. However, it is usual, when calculating, to assume that all cutters set on a block are cutting equally and are all leaving their mark on the surface. The S.I. unit of time is the second, but the minute is acceptable.

The calculation, then, will require that the feeding speed in metres per minute (m/min) is multiplied by 1000 to give feed in millimetres per minute (mm/min). If this figure is divided by the number of cuts occurring each minute, the answer will give the pitch distance of the arc marks, measured in millimetres. Cuts per minute are found by multiplying the number of cutters by the cutter block revolutions per minute (rev/min).

The formula thus constructed will be:

$$P = \frac{1000F}{NR}$$

where F = Feed speed (m/min),
N = Number of cutters set to identical cutting circle,
P = Pitch of cutter mark (mm) and
R = The revolutions per minute of the cutter block (rev/min).

Example 2.1

A cutter block set with four tracked cutters is rotating at 4200 rev/min. The machine is feeding at 24 m/min. Calculate pitch of cutter marks in millimetres.

$$P = \frac{1000F}{NR} = \frac{1000 \times 24}{4 \times 4200} = 1{\cdot}4 \text{ mm}$$

This is very close to 18 cutter marks per inch and would be an acceptable finish for interior-painted joinery.

FEED SPEED

If a graded surface is specified and the machine has a multi-speed gearbox, a similar calculation will give the required feed speed.

$$P = \frac{1000F}{NR} \text{ (as in example 2.1)}$$

Hence, transposing

$$F = \frac{NPR}{1000}$$

to give the formula for F in terms of N, P and R.

Example 2.2

With a cutter block set with two cutters, and that rotates at 4200 rev/min, a specified surface finish of 4 mm pitch is required. At what speed should the gearbox be set?

$$F = \frac{NPR}{1000} = \frac{2 \times 4 \times 4200}{1000} = 33{\cdot}6 \text{ m/min}$$

or as close to this as the machine range will allow.

CUTTING SPEED

The cutting speed for wood planing has usually been close to 5000 ft/min. This takes into consideration the basic dynamics of machines as well as the shearing forces required to rive average wood. Cutter block kinetics, block size limitation, motor standards, etc. are all factors that influence a machine designer when he decides on a cutter block speed. The speed of 5000 ft/min quoted, converts to a reasonable metric figure of 1520 m/min.

The cutting speed of a block is the speed at which the block periphery moves, i.e. the velocity. This is calculated by multiplying the block diameter by π, and then multiplying by the number of revolutions per minute. Then cutter velocity is equal to block peripheral speed. Reasonable units are minutes for time, and metres for distance. As block diameters are given in millimetres (German blocks are stamped ϕ_{mm} meaning literally 'The flying circle diameter') the conversion factor of 1000 is necessary. Peripheral cutting speed P is given by the equation

$$P = \frac{\pi DR}{1000}$$

The block diameter (cutting circle) is fixed by the makers and the only variation possible for a factory engineer lies in the fitting of different diameter pulleys to the block or motor shafts. This is not current practice, but examination syllabuses seem to assume so. Students of machine woodworking are often asked to calculate either the cutting speed of a block, or the revolutions necessary to attain a given 'P' speed.

Using the formula arrived at and transposing to give a formula for rev/min, variation in the set calculations are possible.

$$P = \frac{\pi DR}{1000}$$

Transposing for R

$$R = \frac{1000P}{\pi D}$$

Where D = cutter block diameter (mm), P = cutting speed (m/min), R = number of revolutions per minute and $\pi = 3{\cdot}14$.

Example 2.3

A planing machine is producing work with a poor surface. What is the cutting speed? The block is 150 mm diameter and rotates at 3200 rev/min.

$$P = \frac{\pi DR}{1000} = \frac{3{\cdot}14 \times 150 \times 3200}{1000} = 1507{\cdot}2 \text{ m/min}$$

This is close to the correct speed. Calculate further to obtain the correct speed.

$$R = \frac{1000P}{\pi D} = \frac{1000 \times 1520}{3{\cdot}14 \times 150} = 3227 \text{ rev/min}$$

OUTPUT CALCULATIONS

Examiners sometimes ask for the calculation of machining times. The solution to these is either to assume continuous feed and then to calculate the total run divided by the feed speed, or to work on a basis of interrupted feed at so many pieces per minute.

Example 2.4

How long would it take to surface plane 1·5 m^3 of softwood pieces if each piece had a nominal sawn size of 500 mm × 100 mm × 20 mm? The machinist can average three pieces per minute. Allow 10% extra time for incidental stoppages.

Number of pieces in 1m^3 $= \dfrac{1000 \times 1000 \times 1000}{500 \times 100 \times 20}$

Total number of pieces in 1·5m^3 $= \dfrac{1000 \times 1000 \times 1000 \times 1{\cdot}5}{500 \times 100 \times 20}$

Total time (h) $= \text{Total number of pieces} \times \dfrac{\text{unit time}}{60}$

(Unit time $= \frac{1}{3}$ min)

Total estimated time $= \text{total time} \times \dfrac{110}{100}$

$\therefore$ Total time taken $= \dfrac{1000 \times 1000 \times 1000 \times 1{\cdot}5 \times 110}{500 \times 100 \times 20 \times 3 \times 60 \times 100} = 9{\cdot}16$ h

Power

Power requirements vary with speed of feed, depth of cut, nature of timber and angle of cut. An average allowance of 746 W per 150 mm length of cut is reasonable.

3 Mortising by Machine

The mortise

A mortise is the slot into which the tenon fits to complete a mortise and tenon joint. This joint has a wide variety of forms, many of which, without the aid of special complex machines, are more easily cut by hand. A hand woodworker can introduce the mortise to the wooden base component at any position and angle that his skill allows. Not so the machine, some angles and combinations of length, width and depth defeat it. However, by designing for production and with a knowledge of production methods, most jobs in wood may be machine mortised without loss of strength or appearance; the knowledge of machine potential must be there for this facility to be exercised.

Basically, machine mortising follows the hand process. The hand woodworker marks out the mortise using a gauge set to the chisel he is to use. He then either pre-bores away the bulk of the material between the lines, or he uses his mallet to knock in his mortise chisel to remove waste. If he has pre-bored, this chisel is used only to clean the mortise sides and ends, and this, in essence, is what a machine mortising tool does. The waste, the removal of which forms the mortise, is either chiselled away chip by chip, cut with a series of rapidly passing chisel edges (chain mortising) or is bored away leaving the sides and ends of the mortise to be cleaned by a following chisel edge (hollow square mortise chisels and augers).

One variation of the cutting method has a rapidly rotating cutting bit, which drills into the wood and moves sideways to remove waste (slotting). This method leaves the ends of the mortise rounded. Another variation has a series of sharp teeth ground along an extended chisel which swings into the wood, along and out, rather like a grave digger digging himself into a deepening hole and throwing the removed waste up and out (patent 'Maka' method). Yet another chisel method makes use of two alternating stabbing chisels for mortise ends together with a double edge swinging chisel between, which swings from end to end removing chips severed by the stabbing chisels (patent 'Alternax' method).

Marking for each mortise is not required, only for the initial setting stock and for any mortise that is longer than that that may be cut at one thrust from the machine tool. One-off jobs require marking but for most production runs some form of repetition setting is usually available.

Of these methods, only the chain tool is designed to cut through and out at one stroke. To produce clean entrance and exit holes, all the chisel tools must be

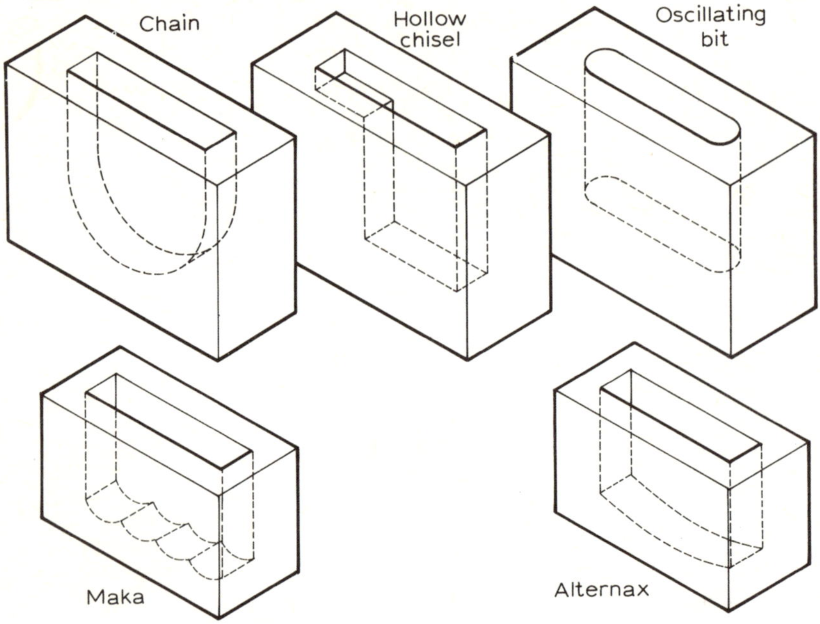

Figure 3.1 Mortise types produced on five different mortising machines

entered from both sides of the wood, or have a close-fitting support strip backing up each piece cut out. The chisel methods all produce reasonably flat bottoms and square ends. The various mortise shapes cut with the tools mentioned are shown in Figure 3.1.

The methods

CHISEL MORTISING

The correct and full name of the tool for this is the *hollow square mortise chisel and auger bit*. With this system there is a hollow and square chisel which has a hole lengthwise throughout its centre. The chisel is set in the machine with one end directed to the wood. The four inside faces of this end are formed to cutting edges.

Internally to the chisel runs an auger (similar in appearance to a hand Jennings bit but without a screw or brad point) that has its two cutting wings larger in diameter than the body of the auger. A centre screw point would pull the auger into the wood and this would not be practical, for feed speed would be difficult to control, and withdrawal would again be most difficult without a reversing motor. A centre, or brad point would serve no purpose as the internal chisel bore holds the auger in line, and a spur would make trouble for the chisel should the operator incautiously allow side feed to the wood whilst the chisel was engaged. Knowledgeable operators use the chisel in this manner to produce clean, flat bottoms to mortises and haunches. The practice is a poor one but it could be claimed that labour costs are more than tool costs and some abuse to the chisel may actually save money.

The auger engages the wood first, and leaves a hole. Immediately following the auger cutting edges are the chisel edges. These cut away the corners left by the auger. The auger diameter being larger than the chisel size (across the flats), the chisel has only very light cutting to do. Chips are fed up the auger flutes and ejected through chip windows in the chisel sides. Side clearance is provided by making the chisel very slightly smaller towards its shank. Figure 3.2 shows this.

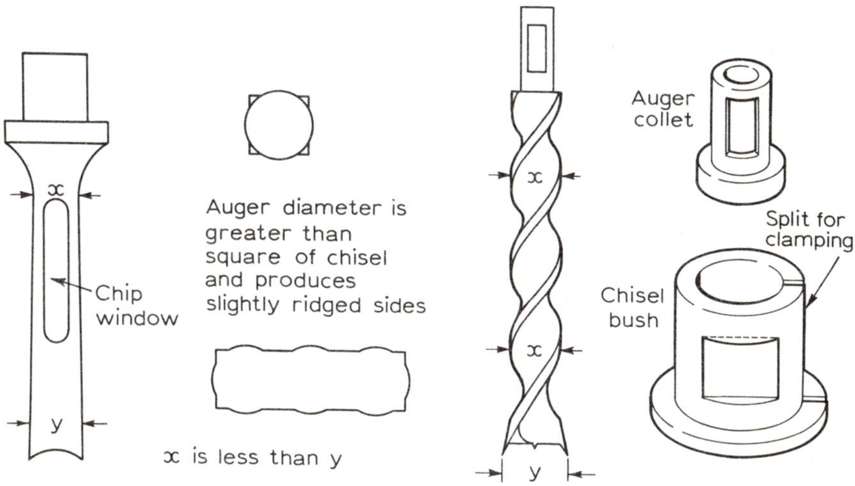

Figure 3.2 Construction of a hollow chisel mortising set. Both chisel and auger are largest at their cutting ends

From this it will be seen that as the chisel is sharpened it suffers a minor decrease in cutting size. Thus machine tenons, which may be infinitely adjusted to size, should be cut after the mortise which may vary according to the age of the chisel or cutting tool used.

At all times, these two fundamentals remain. The auger must lead the chisel into the wood, and the chisel must be smaller across the flats than the cutting diameter of the auger. If a 12 mm size chisel was opened out it would measure 48 mm and hand woodworkers should easily appreciate the forces involved in driving such a flat chisel into the wood. It is the fact that the chisel cuts corners only, that enables the system to work. Figure 3.2 shows this principle and includes a view of the various adaptors used to fit chisel sets to the machine.

Chisel sizes for normal work vary from 6 mm to 25 mm but may go up to 75 mm for heavy coach and lock gate work, but these would be used on special power-driven machines.

Bottom edge break-out is severe with chisel sets unless firm support is given; strictly, only stub mortise and haunches lie within the province of chisel sets, but by reversing wood and cutting from both sides, through mortises may be cut.

Chisel augers run at approximately 3000 rev/min. Power requirements are as follows: up to 1500 W for chisels up to 18 mm, up to 3750 W for chisels up to 32 mm.

CHAIN MORTISING SETS

Mortises cut with chain sets may be through mortises. This is to be preferred, as the chain will always cut a stub mortise with a round bottom, which is not always satisfactory due to the width of material. Figure 3.3 shows the basic features of chain sets. Each set consists of the chain, the guide bar and the driving sprocket. The tool design allows many different mortise widths and

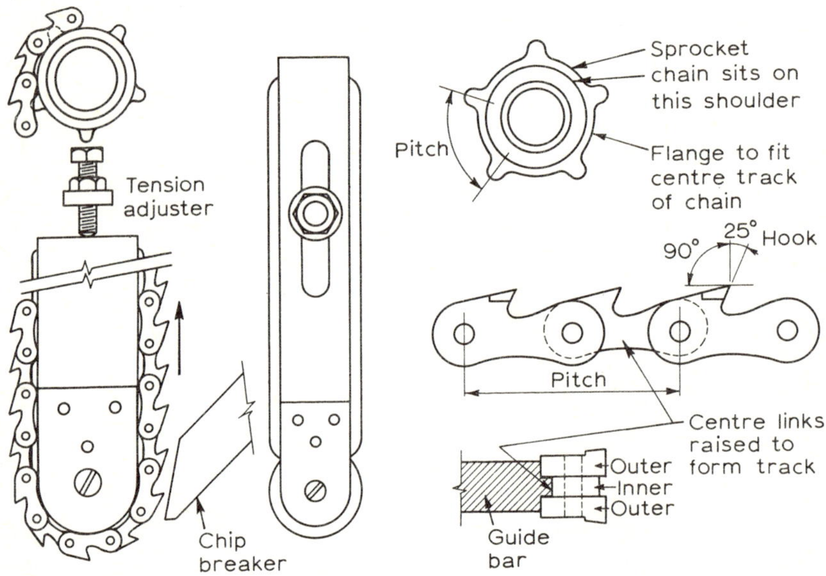

Figure 3.3 Construction principles of a chain and guide bar set

lengths to be cut at one thrust into the wood; each chain and guide bar being matched for this purpose. Longer mortises may be cut by the relative movement of the wood, as with the hollow chisel.

Variations in chain and guide bar combination allow up to 300 different size mortises to be cut at one thrust. The smallest is 4·5 mm × 18 mm and the largest 32 mm × 75 mm. Tapered guide bars allow wedge room to be cut if mortises are cut from the back of the stile. Size variation is possible because chains vary in width from the smallest (4·5 mm) to the largest (32 mm) going up in stages of approximately 1·5 mm. The guide bar also goes up in thickness, but in double steps, so there are less guide bar sizes than chains. Guide bars also have standard width increases, for it is the width of the guide bar plus two thicknesses of the chain that give the mortise lengths.

Pitch size is the measure of distances between the teeth of two following side plates. Chains are made in three pitch sizes, i.e. 0·54 in (13·4 mm), 0·62 in (15·74 mm) and 0·89 in (22·6 mm). However, it should be noted that mortising tools, chisels and chain sets are not, at time of writing, available in metric sizes, and the figures and values given here are Imperial to S.I. *conversions*, and are thus only approximately correct. New standards will eventually be adopted.

The mortise length that may be cut by any chain set combination at one thrust into wood is determined by the width of the guide bar chosen and two thicknesses of the chain. The chain thickness will be smaller with the smaller pitch. To enable narrow guide bars to produce short mortises, light chains are used which have only narrow thickness. At the lower end of the guide bar there is an anti-friction roller which conducts the running chain through an arc of 180°. To negotiate the roller of a narrow guide bar the chain links must be short in length to avoid see-sawing, which would occur if the links were too long. Thus light chains for short mortises have a short pitch. As the mortise length increases heavier chains are desirable to allow full exploitation of the fast speed at which chain sets can cut. These larger mortises are cut with wider guide bars having larger diameter anti-friction rollers which allow longer chain pitches to negotiate them. As has been stated, the three pitches are 13·4 mm, 15·74 mm and 22·6 mm. Chains are made up of three-, five- or seven-link plates in width, and are all riveted through. Figure 3.4 shows the construction of a five- and three-plate chain together with guide-bar assembly.

Each outer plate has a slight projection that gives clearance in cut in a similar manner to a carbide-tipped saw tooth. The centre plates are reduced at their cutting edges to half of their width. This relieves the shock of the cut, and following links have this width reduction alternating from side to side.

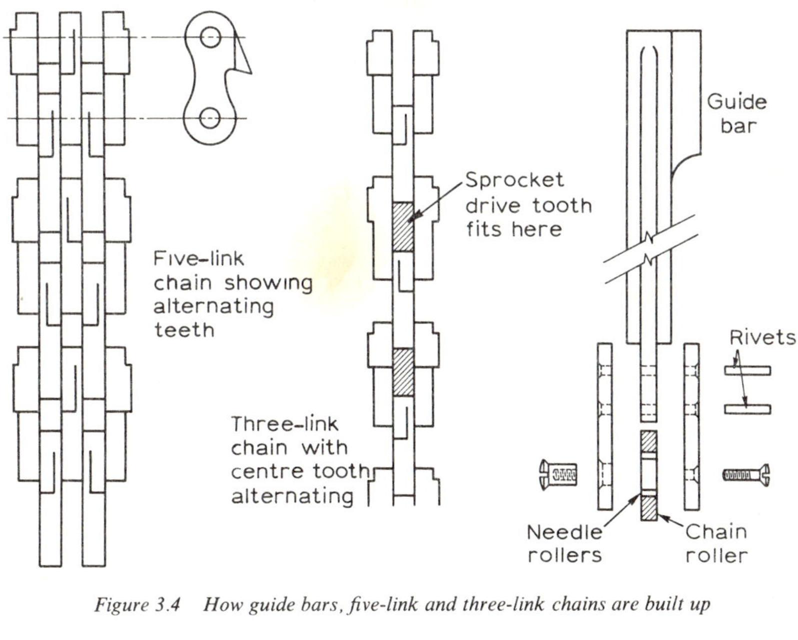

Figure 3.4 How guide bars, five-link and three-link chains are built up

As the chain cutting teeth emerge from the wood, lifting of grain fibres occurs. This is minimised by careful setting of the wooden chip-breaking block, set close to the upward side of the chain. The chip breaker is impressed into the top face of the wood by a heavy weight bolted to the rod on which the chip breaker is fixed.

To allow for free running and minimum wear, chain sets must be run slack and frequently oiled. When setting, the chain must be left no tighter than will allow a pencil to be easily pushed between chain and guide bar. This allows for the heat expansion due to friction that lengthens the guide bar during use. Chain sets run at approximately 3000 rev/min, and require power of 1500–3000 W.

THE SLOTTER

The correct name is the *oscillating bit mortiser*, but in use it is often called the *slotter* or the *jigger*. The cutting bit is a double-fluted fast-rotating end-cutting tool. The flutes terminate in two sharp cutting teeth. These are the largest diameter, and bits, like square chisels, cut very slightly less after each sharpening.

As the bit rotates it oscillates from side to side, not in a wig-wagging way but in a rolling way, with its whole length moving. As this oscillation takes place the bit is fed into the wood, or the wood to the bit, and the two end points cut away a channel of increasing depth. This forms the mortising slot, which has a flat bottom and rounded ends. Slotters are intended for stubbed mortises, but may cut through if the back- or break-out point is supported. Some slotting bits have the edge of one flute serrated into cutting teeth with the intent to masticate the cut chips to facilitate removal, as unlike chains and chisels there is no natural chip removal. Figure 3.5 shows a typical slotting bit and indicates the method of

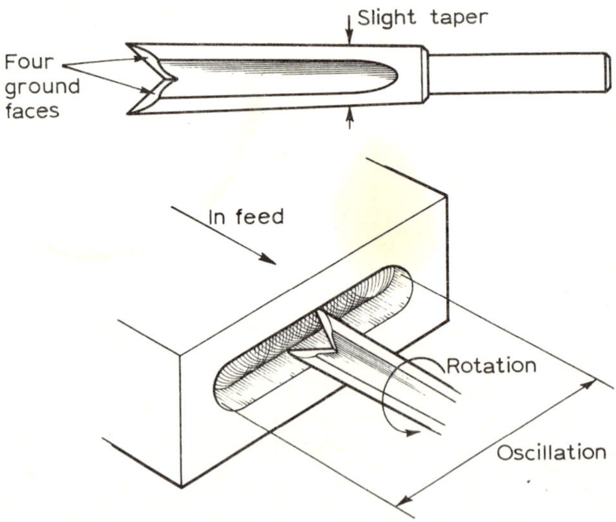

Figure 3.5 Slotting bit. This cuts on its two end points which are its largest diameter, unlike a router cutter that cuts on its side

work. Bits rotate at approximately 8500 rev/min. Power required is 1100–1500 W.

THE MAKA

The patent *Maka* system is based on a swing chisel that enters wood, swings and lifts away, cutting and clearing chips as it goes. The mortise bottom (normally all are stub mortises) is comparatively flat, and is made up of a series of large ripple marks rather like a very much enlarged rotary planed surface.

The design of the tool and of the cutting system is based on the geometrical principles of loci—the paths of moving points. The size mortise cut, the tool selected and the setting of the machine moving parts are all closely knit together. Each factor must be correctly and closely related to the other.

At the top, or non-cutting end of the tool, there is an eccentric shaft. This projects through a circular disc which moves a beam on which the cutter is bolted. As the eccentric shaft rotates, it raises and lowers the cutting tool as well as moving it from side to side. The circular disc which controls the cutting tool is free to move in this way, but is itself housed in a beam casting that is pivoted above the eccentric shaft. Thus, as the sideways and up-and-down movement of the tool occurs, it cannot be a circular movement but must be elliptical, and the chisel end swings from side to side, with the major axis of the elliptical path being parallel to the length of the mortise. Figure 3.6*a* shows the geometrical principles involved.

The tool itself has a long narrow shank. The cutting end is wider and thicker to give clearance. There are two types of teeth on the smaller sizes, and three or sometimes four on the larger. On the smaller tools having only two teeth, one is larger than the other. The larger one, which is the leader, is chisel edged and approaches the wood at a flat angle. This is what cuts the chips. Following this there is another chisel edge that points directly down into the wood. The job of this one is to sever the rear end of the chip and keep one end of the mortise square and sharp. At the other end of the mortise the wood is supported by a wooden chip-breaking pad that holds the fibres as the leading chisel edge emerges from the wood and breaks-off each chip. On larger tools, the chisel-edge chip-breaker tooth is followed by one or more small teeth that help to keep the bottom of the mortise flat. These are always followed by the vertical end cutter. The leading edge of the tool may be serrated to assist fracture of the chips and to assist in raising them from the deepening mortise. Figure 3.6*b* shows details of the Maka chisel tools.

By rotating the eccentric within the tool holding disc, shorter or longer strokes are made possible, thus shortening or lengthening the cut mortise. Each tool has only a limited facility to cover extended strokes so that tools must be changed if a radical change of mortise length is required.

As with all rotating bodies, balance is important. All forces acting about the pivot centre must, as near as possible, be opposite and equal pairs. If a point in space is subjected to pulling forces in pairs, all being opposite and equal, the point stays still. This is the situation machine woodworkers endeavour to create on the centre line of all rotating shafts. It is only by doing this that running

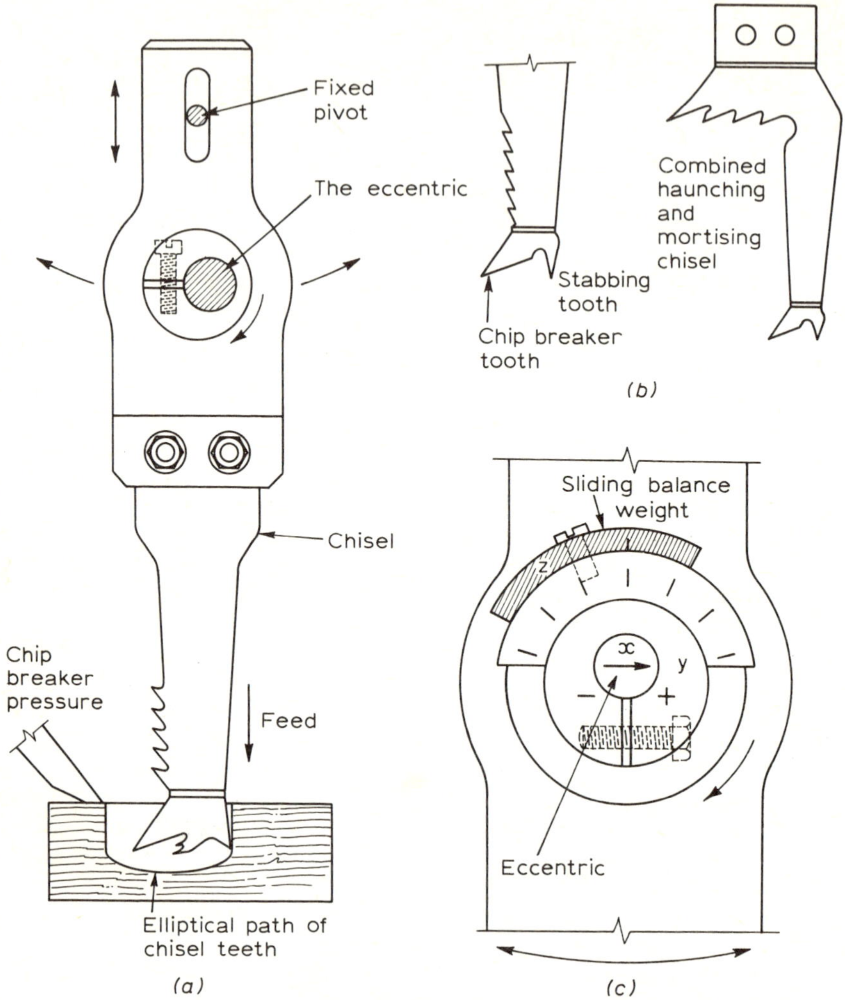

Figure 3.6 The Maka system. (a) *Geometrical principles,* (b) *typical Maka chisels and* (c) *balance mechanism. Rotation of the eccentric shaft* x *within disc* y *to alter the mortise length necessitates adjustment of weight* z *to maintain running balance*

balance may be achieved. However, all mechanisms vary, and fine balancing is difficult and at the best is a compromise. This is the case with the Maka. A tool of given length will cut a maximum or a minimum length mortise controlled by the setting of the eccentric.

The cutter control disc has a sliding weight attached by screws. Adjustment of this compensates for the varying degrees of eccentric rotation called for to alter mortise length. Figure 3.6*c* shows the Maka system of balance.

Tool adjustment and selection at first appears complex, but the manufacturers supply a handbook which provides charts and guide lines for the information of the operator.

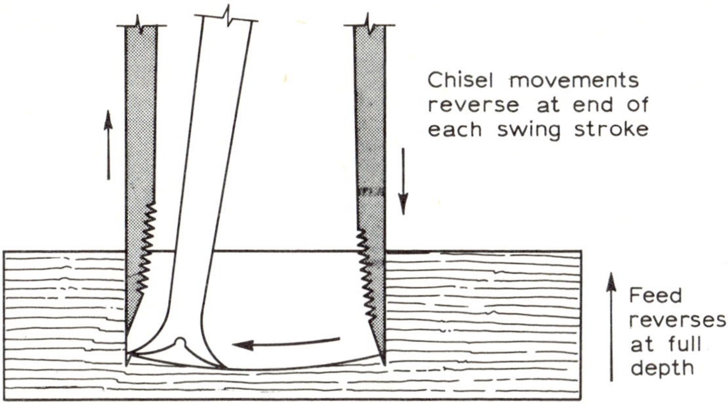

Figure 3.7 The Alternax system with a swing chisel leaves a slightly round bottom to the mortise. The alternating stabbing chisels are also shown

Mortise sizes may be as narrow as 1·5 mm and as short as 12 mm, or as wide as 18 mm and as long as 100 mm. Speed of oscillation is approximately 3000 rev/min, and power required varies from 1100 to 1500 W. Hand or air clamping is available and the machine may be pivoted over 90° to make possible mortising for door locks, etc. after assembly.

THE ALTERNAX

The patent name *Alternax* springs from the cutting system. The mortise is divided into two ends, and a centre, waste portion. There are three chisels, two of which cut alternately, and a centre one that swings from end to end of the mortise. Cutting action is shown in Figure 3.7.

The System

One chisel, shaped like a conventional hand mortise chisel but serrated on its back face above the ground bevel, stabs outwards from the machine and cuts a clean cut line to form one end of the mortise; it then retracts. The centre chisel, which has two edges, one hooked to each end of the mortise, then swings towards the stabbed cut and in doing so removes a chip. At the same time, the chisel that is to form the other end of the mortise, stabs outwards in its turn. As this second stabbing chisel retracts, the swing chisel reverses and removes a chip at the second end. Whilst this is occurring, the first chisel stabs again, and so the cycle is repeated. Thus the mortise is cut with clean square ends and an approximately flat bottom. With suitable back support, through mortises can be cut. Bevel-sided chisels enable dove-tailed housings and bevel-scribed tenons to be cut.

The full machine cycle is automatic once the initiating signal is given by hand lever or foot valve. The air clamp goes on first, and then the table (with wood) is fed into the swinging chisels. At the completion of the cycle the table is at rest and retracted from the chisels.

Adjustments for mortise lengths are made by hand wheel, and can be done while the machine is in motion. The mechanical principle is again that of a variable eccentric, and the hand wheel simply increases or decreases the degree of eccentricity. Mortise sizes vary from a minimum of 24 mm long to 85 mm; depth of cut is up to 85 mm. To change the width of mortise the three chisels must be changed and sizes available range from 3 mm to 18 mm. There are approximately 1500 full cutting strokes per minute. Power required is 1100 W.

Tool maintenance

CHISELS AND AUGERS

Chisels

Chisels are sharpened by grinding or by the use of reamers. Final sharpening is done by hand filing.

The insides of the chisel edges have two bevels (Figure 3.8). The first of these two bevels, looking up the chisel, is the sharpness bevel of 35°–40°. The second is the clearance bevel and is at approximately 25°. This sharpness bevel is formed

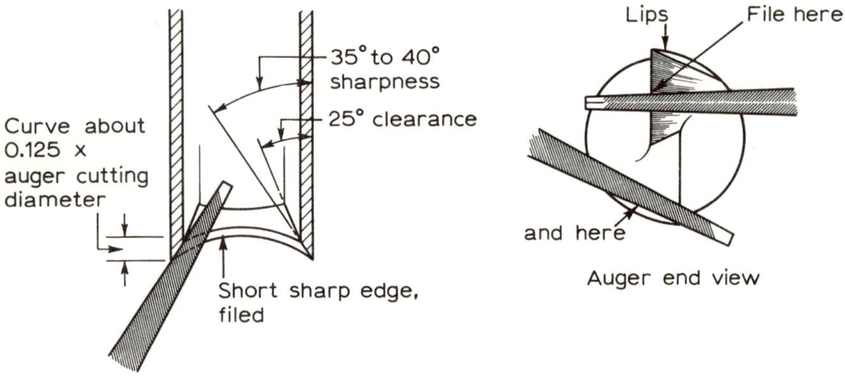

Figure 3.8 Hollow chisel maintenance. Chisels have a sharp edge of 35° which is relieved another 10° for chip clearance. Both chisel and auger have their edges filed for final sharpness

using either a hand-applied conical hard-steel reamer, or by using a conical grinding wheel. The reamer, which has a pilot bush to fit the chisel centre hole, is probably the best tool, as unskilled use of the grinding wheel leads to grooves and shoulders, and thus blunts chisels. The reamer, however, if used too heavily, will produce a substantial burr.

The second bevel is usually filed and is made broad enough to reach out to the internal corners. Too heavy filing at corners will cause chisels to split. Some operators decrease the sharpness bevel and rely on this to give sufficient clearance.

The final condition of the chisel should be that all edges are sharp and level, and that the curve between points is distinct and sufficient to clear the rotating wings of the auger. Inevitably, a burr will be created on the outside of the chisel.

For examination answers this is ignored; the 'book' says 'Thou shalt not touch the outside face of this chisel'. However, in most cases the burr is there. A very light application of the chisel to a fine flat oilstone will remove it and leave a sharp edge. The danger lies in the fact that chisels are tapered into their shank ends for clearance, and any careless rubbing of the chisel outer face would remove this clearance and make chisel penetration difficult.

Augers

Augers are filed lightly on the insides of the scribing wings and on the undersides of the cutting edges. The circle scribed by the lips is larger than the square of the chisel and must be maintained so, for the smaller the hole cut by the auger, the heavier will be the work of the chisel. The flutes of the auger should be lightly polished with fine emery cloth from time to time to remove slight burrs that may impede chip removal.

CHAIN SETS

Chains

Chain maintenance requires chain grinding, chain-link repair and lubrication. Most mortise machines have a built-on attachment on which chains are ground. This attachment consists of a grinding wheel and arbor, a sliding collar fitted with sprockets to carry the chain, a ratchet device to index each tooth, and two screw adjustments (up and down, and horizontal) for the purpose of the alignment of chain to wheel, and for compensating for wheel wear.

The wheel may be flat or bevelled, and range from 75 mm to 125 mm diameter. Its specification will vary to choice. Mostly, the abrasive is aluminium oxide with a grit size of 46 to 60. Bonding is mainly vitreous, although for safety with such thin wheels, resinoid or rubber bonds are used. Bond hardness is in H/K range.

Rubber-bonded wheels cut very smoothly and are almost impervious to damage should the chain jump a sprocket when being ground, but they create a smell of burnt rubber and are unpleasant to use. Figure 3.9*a* shows the relationship between chain tooth, wheel, sprocket and ratchet. Details vary, but in general these sprocket collars have facilities for one pitch chain only and must be changed if another pitch is to be ground. They either have multiple sprockets of varying thickness to cater for narrow or wide chains or they have pairs of thin sprockets that may be adjusted to each chain width.

Having selected the correct sprocket the adjustments are arranged until the ratchet device aligns the chain tooth correctly with the wheel face to give a grinding angle of 25°. This is the 'hook' angle, and is the angle at which the tooth face lies in relationship to the line of the chain back. Chains may have more, or less hook, but if the hook is too great chains will snatch and spoil ends of mortises, whilst hook of too small a degree will lead to choked teeth and slow feeding.

Having aligned the chain, the sprocket collar is moved freely by hand under the wheel causing light contact between wheel and chain tooth face. The chain

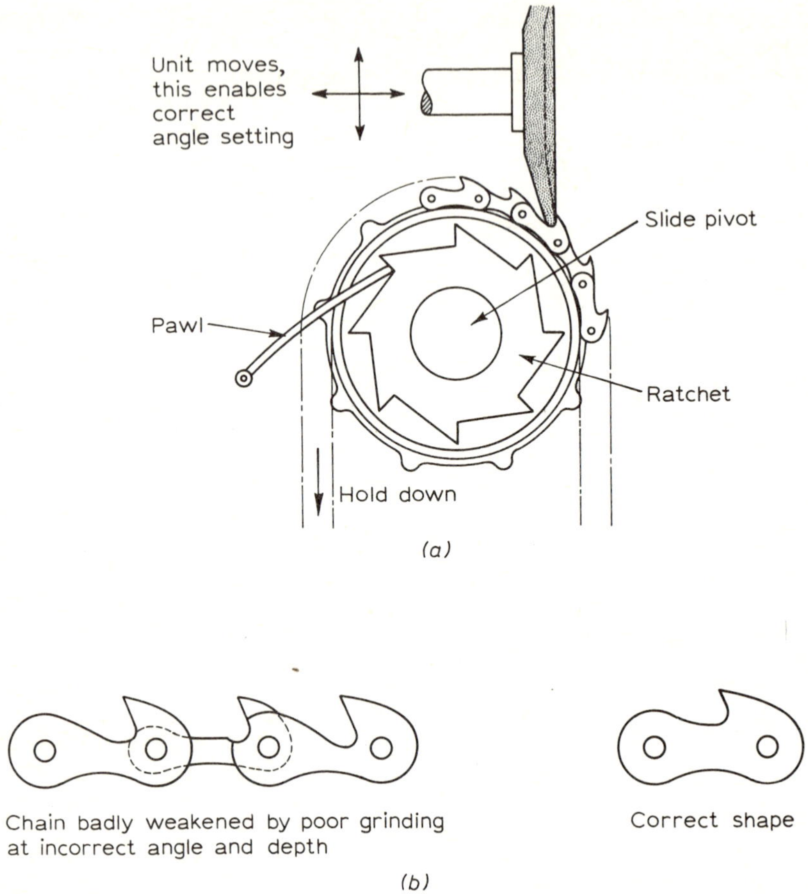

Figure 3.9 Chain grinding. (a) *Chains being ground sit on a sprocket carried on a sliding spool, which is ratchet controlled; the grinding unit has a full range of movement to enable chains to be aligned at the correct angle of 25°.* (b) *Overgrinding causes broken chains*

is pulled down all the time to ensure a snug fit between it and the sprocket. Between grinds the chain may be slipstoned on the faces of the teeth. This takes time and is very rarely done. Chains may be stored flat in an oil bath or hung into a can of oil. Thinking ahead is necessary so that the next chain in use may be hung to drip-dry before use.

GRINDING FAULTS

Too heavy cutting by the wheel will burn the chains, which, being of high-carbon steel, will then lose their ability to keep a sharp edge. Poor fitting of the chain to the sprocket will allow it to be ground out of square, and will cause it to run to one side when in use. Gullets that are too deeply ground weaken the chain and eventually the grinding wheel will approach a rivet. Then, in use, the chain will break.

Hence the rule is light grinding only, to the correct depth and angle.

CHAIN REPAIRS

A wide chain is complex, having up to seven lines of side plates (links) riveted together (Figure 3.4). Figure 3.9*b* shows the dangerous nature of a chain that has been ground too close to rivets.

The repair of a chain is reasonably simple, but requires the correct set of tools for each chain. These tools are the 'first breaking punch', which punches the rivet through sufficiently for the 'second breaking punch' to fit into the hole and drive the rivet right through and out. The chain is supported for this on an 'anvil' which is complex and of hard steel. It has a wide groove on its end for the purpose of breaking the chain (allowing two rivets and one side plate to be punched out and off). The anvil also has two pairs of holes, one pair of which is deeper than the longest rivets used so that old rivets may be punched into them thus freeing the side plate. The other pair of holes is slightly less in depth than the shortest rivets used. These allow support while the new side plate is being tapped on a pair of rivets. Along the sides of the anvil are two grooves, shaped to support wide or narrow chains while the final riveting is being done with the 'making punch' which spreads rivet tops into counter-sunk holes in the side plates.

The repair process is summarised as follows:

1. Grind freehand to remove rivet heads.
2. Support the chain on the anvil end groove and break the rivet grip with the first punch. This drives out two rivets and one side plate.
3. Use the parallel second punch and drive right through, pushing out one pair of rivets and one side plate.
4. Drop the rivets into the deep pair of holes and drive free of the side plate.
5. Select new rivets/side plates/link plates, as required, and fit two rivets into the shallow anvil holes. Knock on the correct side plate. Assemble the chain to the correct width.
6. Drop the complete chain into the correct side groove and use the making punch to snap rivets into counter-sunk holes.
7. Freehand grind excess rivet heads.
8. Freehand grind new chain links down to the line of the old ones.
9. Lubricate thoroughly and run on the machine for a few minutes without cutting.

Figure 3.10 shows the basic details of the chain repair process.

GUIDE BARS

Most guide bars have repairable bearings for their anti-friction rollers. The general question is whether the high cost of today's labour justifies the saving in price of a new guide bar. Guide bars, although not cheap, give years of service. However, if a repair is to be made the rivets holding the roller side plate

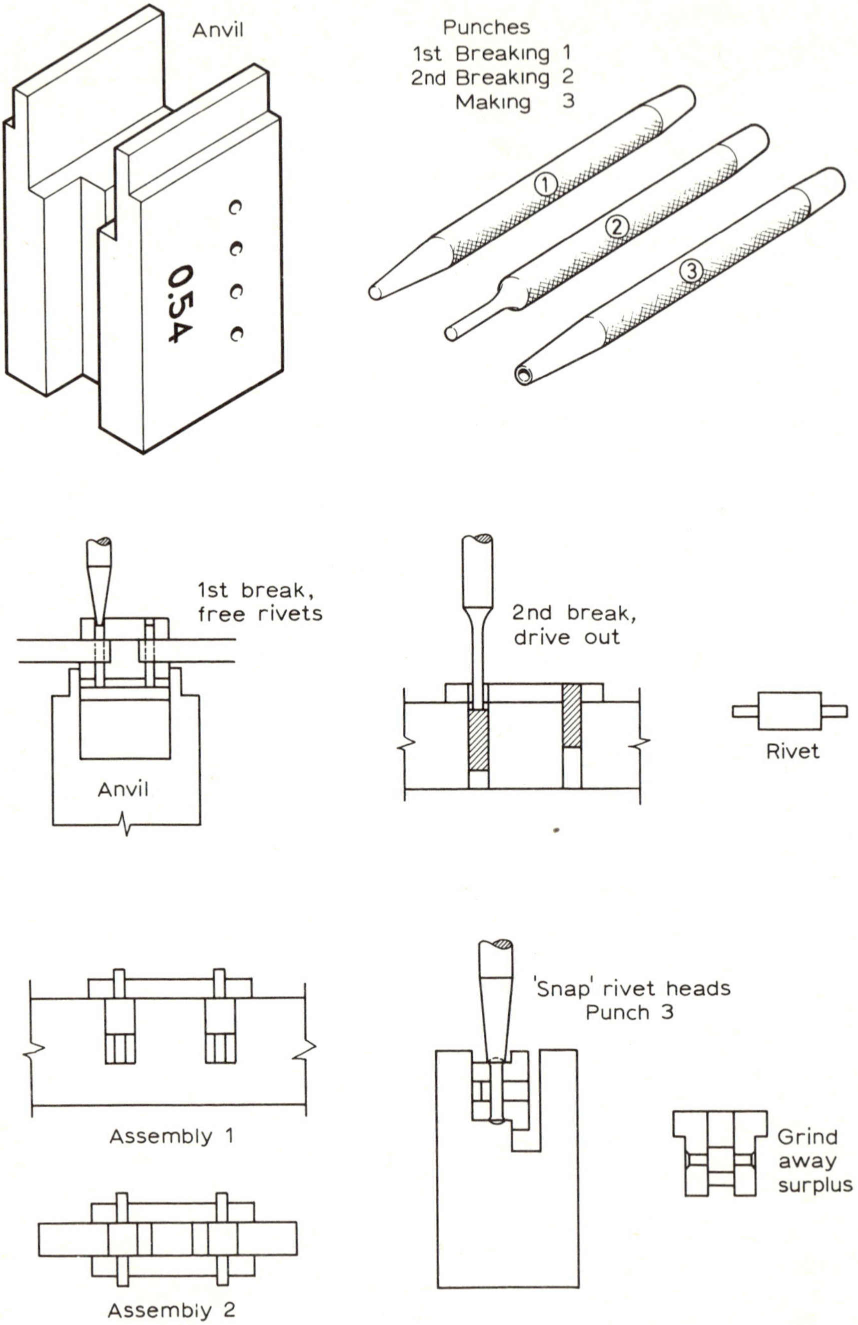

Figure 3.10 *Chain repairs. Broken chains can be repaired; there must be a repair set for each pitch of chain, and the procedure shown here is usually followed*

are driven out and the centre screw is removed. Old wheel and needle rollers are then removed and the new set, needles, roller and centre collar, are greased into place. The side plate is replaced, riveted and screwed. Rivets must be filed flat. This completes the repair.

SLOTTING BITS

These are mostly ground free-hand. They can be machine ground, but as this usually means a complex set-up on the universal grinder, it is not often practised. Like so much in machine woodworking, the original angles are followed until the particular tool ceases to cut well, then it is altered until it cuts well again.

There are four bevels that need to be ground. There is one pair to each cutting point, and the object is to achieve two sharp, strong, points, level with each other, each having sufficient (but no more) clearance behind cutting points. The included internal angle is kept at approximately 70°. An occasional polish of the flutes and outside the cutters is all that should be needed.

MAKA TOOLS

These are high-speed steel and require grinding with white aluminium oxide wheels of 60 grit (a few tools are tipped with tungsten carbide for abrasive timbers, and these must be ground with green silicon carbide and lapped with diamond wheels). The angles at which the teeth are ground are critical and involved, and the machine handbook must be followed.

In general, the leading chip-breaking tooth *must* lead. It will have an approach angle of 35°, a sharpness angle of 40° and a clearance angle of 15°. The cutting teeth have a hook of 5°, and the final stabbing tooth must be slender, and be very slightly leaning forward towards the direction of the cut. The instruction manual gives an angle of 89° for this inclination. If this forward inclination is too great the tooth will wedge and not cut.

The finished tool must have all teeth accurately level and square across, with deep gullets. Even slight variation from the precise specification will cause Maka tools to drag, and produce a poor mortise.

THE ALTERNAX

The chisels here are simple and require only commonsense sharpening similar to hand chisels, i.e. grind when the edge is thick and oilstone sharpen between grinds, and keep strong, sharp edges with sufficient clearance on the swinging two-edged chisel, paying close attention to clearance between the sole of the chisel and the wood face.

Faults in mortising

The most common faults in mortising are shown in Figure 3.11. Chains that are run too slack will round entrance and exit ends of the mortise, as they wrap round before being forced inwards to the guide bar. Guide bars with worn rollers will

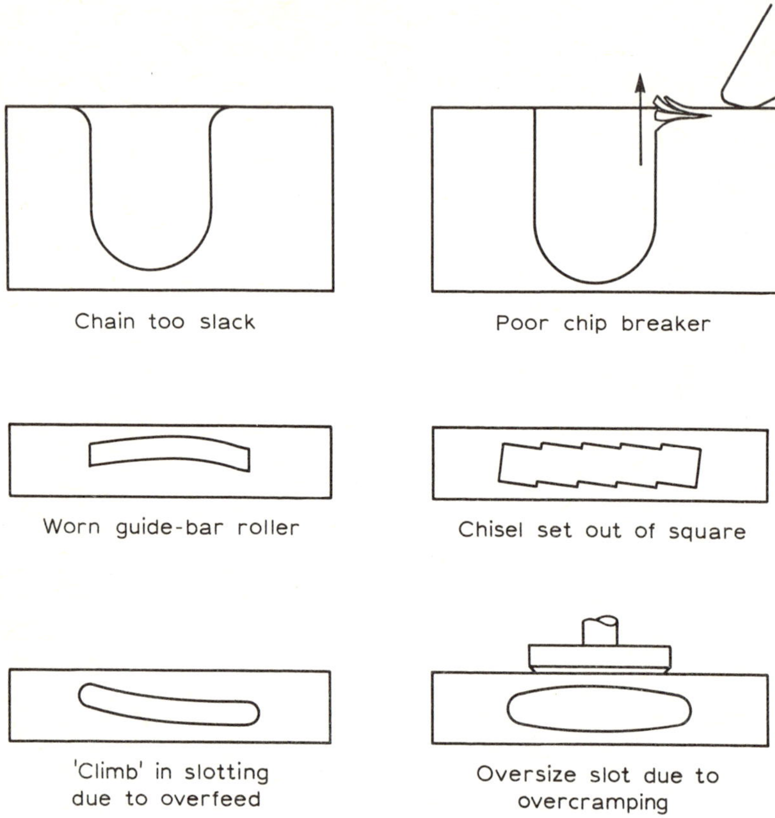

Figure 3.11 Faults in mortising. Slack chains cause rounded ends; broken ends are due to faulty chips breaking; curved mortises are due to the guide-bar roller canting sideways; stepped edges from the chisel indicate out-of-square setting; cramping above the mortise compresses the wood which recovers when released

cause crescent-shaped mortises, as the extreme lower end of the chain is forced over by inclination of the roller.

Chisels split at lower ends due to a number of different reasons. These include too heavy feeding; too heavy internal filing which has caused thin walls; too little clearance between chisel edges and auger lips, causing overheating and weakening; too small a diameter of auger lips causing extra large chips to be forced up into the chisel; augers set too far out, causing large chips that can split the chisel. Zig-zag edges to a mortise cut with a square chisel indicate out-of-square setting.

Slotting bits will climb, i.e. run out of parallel, if they have cutting points of unequal diameter or length. Choking due to too heavy oscillation feed or too heavy in-feed will also cause climb.

Woolly mortises from Maka chisels indicate incorrect sharpening, blunt tools or a poor relationship between chisel size, length of mortise and stroke length. Noise from the Maka is a sure sign of incorrectly set balance weights.

ALTERNAX

Alternax faults are few. Apart from blunt chisels, the only other failing can be incorrect setting, with stabbing chisels set too far out or too far back in relationship to the swinging chisel. If a stabbing chisel is too far out it will leave a sharp chisel cut at the end of the mortise bottom, and if too far back it will leave a torn end when the swinging chisel is forced to wrench loose an imperfectly separated chip. If chisels are kept in sets and all ground equally, no difficulty should occur.

The machines

Mortise machines are either single or 'gang' (multiple) and may be vertical or horizontal. They may be hand or power applied, and in the latter case the power may be mechanical, pneumatic or hydro-pneumatic.

The simple chain and chisel machines are vertical, and most have a hand feed by means of a long lever. Apart from operation, the machine man must know that the weight of the cutting head is counterbalanced by either a weight or a coil spring concealed within the machine. Only the spring type can be adjusted and should be set so that there is a slight tendency for the head to lift itself as the hand lever is raised. On no account must the machine head be inclined to fall under its own weight.

The hand lever usually has a pivot casting that permits adjustment. This enables each operator to find the best length and angle of lever to suit the length and cutting depth of the chain or chisel he is using. Too many operators accept the lever as it is and never re-set to suit work conditions.

Chain mortises are never horizontal because of the fall of slack chain. They are, however, made in gang form with up to eight chains in a row. To achieve really high production, chains may be set in pairs, one in front of the other, with spacers in between. Thus a pair of door stiles can be cut at one pass.

Gang chisels are usually horizontal, with the work table moving in towards the ends of the chisel. The work table has transverse slides and its position is controlled by a large hand wheel as with the single, vertical mortises.

One form of vertical gang mortiser, the *Brookman* has three cutting heads, chain or chisel. A unique control system called *Hydraulectric* enables a wide range of operational procedures. Head feed and work clamps are automatic, but are subject to overriding control by the operator's foot pedal.

The principle of control is that of oil flow through restrictor valves, with the oil being pumped by an electric motor. Fine setting enables the cutting heads to have a fast approach and retraction, but controlled cutting feed to suit size of tools and material (hard or soft wood). The head cutting stroke is controlled by a spring, which, compressed by hydraulic power, acts as a power accumulator. Once control of the machine has been mastered, operators manage to achieve sufficient up-and-down strokes of the cutting head to cut an average length mortise (using chisels) with only one movement of the foot. At all times the machine circuit must be free of air. Spongy head movement indicates air in the power circuit, and failure of work clamps or work hold-downs is due to the same cause, i.e. air in circuit. Standard bleeding procedure must be followed until no air bubbles are present.

Multiple Maka mortises are usually horizontal and have up to six heads. Work is held by air clamps and table movement is inwards to the chisels, controlled again by air. The essence of this type of machine is that each work piece must be fully mortised by correct selection and setting of chisels, at one inward feeding stroke. A stile having a narrow, haunched top-rail mortise, three glass bar mortises and a double, haunched, bottom-rail mortise, could be cut at one stroke, leaving only one last haunch to be cut later.

The slotting mortiser is made as a fully-automatic five-headed double-sided machine, and in several forms as a single hand-operated or single automatic machine. Three-headed machines are also made. All are horizontal.

The fully-automatic five-headed machine has a continuous working cycle. The operator stands at one end, where all controls are situated, and feeds first to the left and then to the right. The cutting heads are central to the machine and have double-ended chucks holding slotting bits driven by short-centre 'V' belts. The bits rotate and oscillate; feed is by sliding work tables that carry the work into the cutting bits. The two work tables may be worked as a pair. As one is in-feeding, the other is out-feeding. On the work tables there are clamps and the various jigs that are required to present shaped and straight workpieces to the cutters at the correct angle and position.

4 Tenoning

The tenon fit

To cut a *tenon* is to produce a spigot or male member of a mortise and tenon joint. The tenon fits into the mortise and it is the degree of accuracy with which it fits that determines the strength of the joint. A poor fit gives poor strength, and gap-filling adhesives cannot alter this. Gap fillers will overcome discrepancies in fit where close fitting and pressing are not possible, and in such situations they produce the strongest joint possible, but they are not an excuse for poor fitting. The fit of the joint is related to the tenon and not to the mortise, for whereas mortising tools are made to a nominal size and cannot easily be varied, tenoning machines have infinite adjustment for tenon size.

Experience and knowledge of the job requirements are the only guide to what is, and what is not, a good fit. A small tenon for carcase furniture in hardwood is a good fit if a smart tap with a hammer or mallet will drive it home. The same degree of tightness of tenon joint in a heavy prestige type of entrance door would render the joint uncloseable by any hammer, including the sledge. A *fit* is that degree of tightness which just permits the closing of the glued joint under reasonable applied pressure. A reasonable rule of thumb standard is that where a corner of the tenon is pushed home by hand it should support the weight of the rail as it hangs. This test produces tenons that are slightly loose for knock-in framing but slightly tight for hand-finished hard-wood joinery.

Overall sizes for sections, and cut-off lengths and shoulder lengths are reasonably easily fitted into a Go and No Go gauging system, whereas tenoning fits are not. Timber responds to cutting according to its nature and to its moisture content, cutters wear during the course of production of long runs and mortised rails shrink before assembly. All these factors make standard 'fits' difficult to apply to tenoned joints. Figure 4.1 shows a selection of tenon types that can be cut on machine tenoners.

Tenoning methods

Tenons can be cut on several machines other than standard tenoners. Twin heads separated by spacing collars, used on spindle moulders and with the wood end passed between them, produce good tenons. Alternate cheek cuts can be made on routing machines, and larger tenons can be cut on a circular-saw bench using

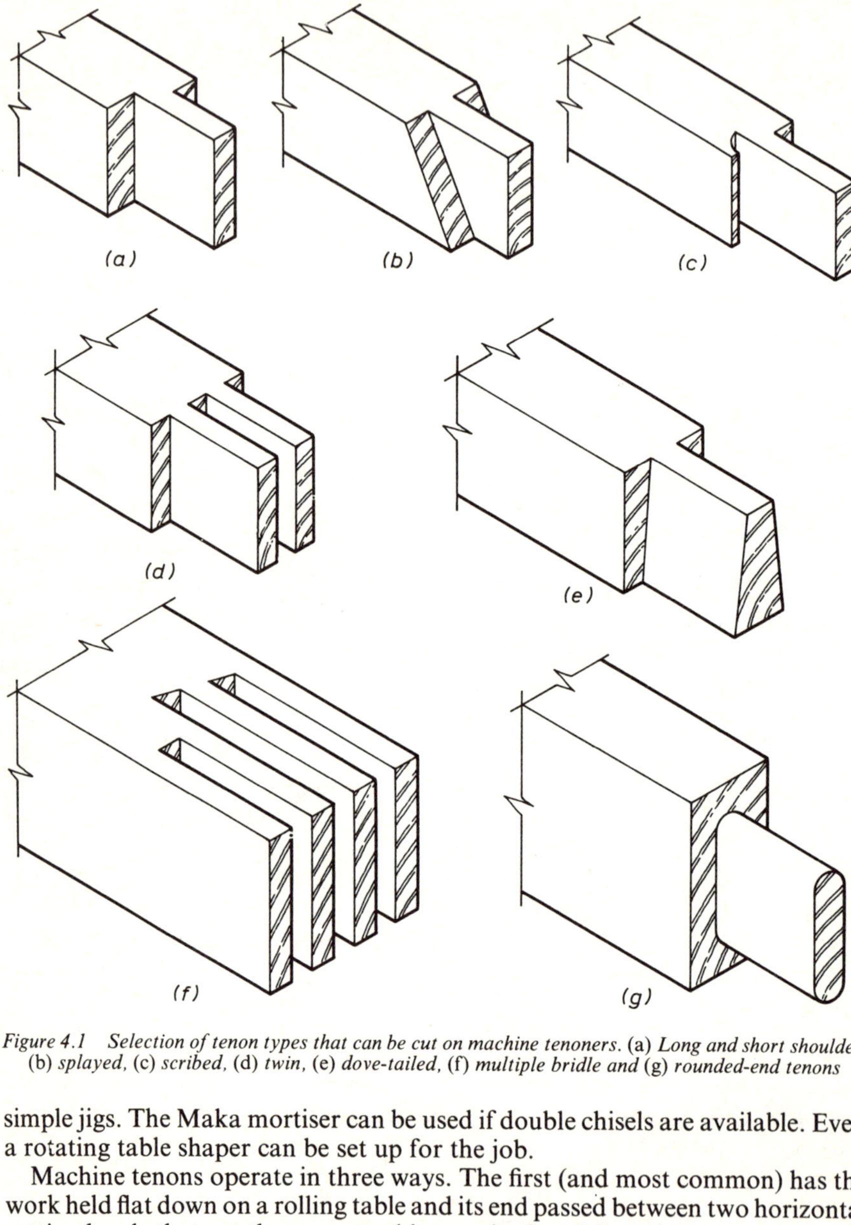

Figure 4.1 *Selection of tenon types that can be cut on machine tenoners.* (a) *Long and short shoulder,* (b) *splayed,* (c) *scribed,* (d) *twin,* (e) *dove-tailed,* (f) *multiple bridle and* (g) *rounded-end tenons*

simple jigs. The Maka mortiser can be used if double chisels are available. Even a rotating table shaper can be set up for the job.

Machine tenons operate in three ways. The first (and most common) has the work held flat down on a rolling table and its end passed between two horizontal cutting heads that cut the upper and lower cheeks of the tenon.

Copying the hand method, there are cutters that incise lines across the grain (as with the use of the marking knife), and these are followed by chisel acting cutters that pare away the cheek until the required thickness is left. The rail is then taken from the machine. An alternative design has two cutter heads running end-on to each other at the top and bottom of two vertical motor shafts. The

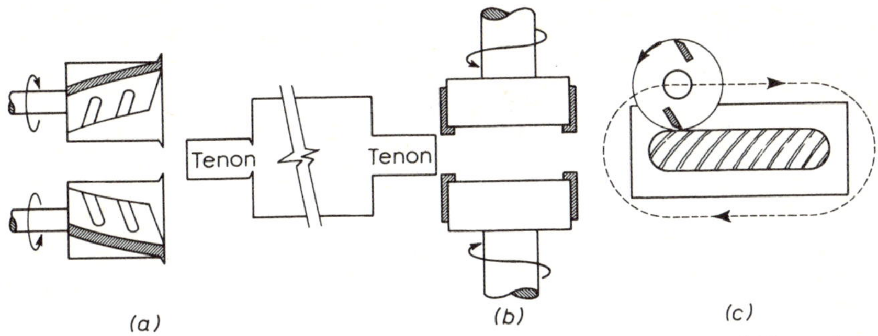

Figure 4.2 Cutting methods. (a) *Horizontal cutter blocks,* (b) *vertical blocks and* (c) *rounded end*

rail is again passed between these heads and the cheeks are cut away. There are no incised lines across the grain at the shoulder, for the cutters cut end on to the grain and have no tendency to lift it.

The third machine form is for the production of tenons having rounded sides. (The sides of tenons fit to the ends of mortises, so most often round-sided tenons are called *rounded-end* tenons, which is not strictly correct.) These tenons are achieved by one cutter block, carrying shoulder spurs and chisel cutters. The cutter block moves round the end of the rail in a path that leaves the tenon the required width and thickness and with two rounded sides (called *ends*).

These cutting principles are shown in Figure 4.2.

The machines

Tenoning machines may be hand applied or automatic in action, requiring only that they are fed. They may be single ended, cutting one end of a rail at a time, or they may be double ended, cutting both ends at once. The larger varieties of double enders have continuous feed and accept work at one end and jettison it at the other. These machines are complex and perform many additional and alternative operations to that of simple tenoning.

THE STANDARD SINGLE-ENDED TENONER

On an 'L'-shaped base casting (with the longer arm vertical) there are two tenoning heads, a cut-off saw and two scribing heads. A minimum of two tenoning heads may be fitted, with the others as optional extras. There is also the work-carrying rolling table which itself carries a guide fence, work clamp and shoulder-length stops.

It is usual to speak of the cutter blocks on tenoning as *heads*. The work process is simple; the heads are manually adjusted (set) and when correctly positioned (proved by trial cuts) the work is placed face down on the rolling table, pushed back to the fence, along to the end stop and then clamped. The table is then rolled forward (away from operator), thus carrying the overhanging rail end into the orbit of the rotating cutter heads. Material is cut away chip by chip and

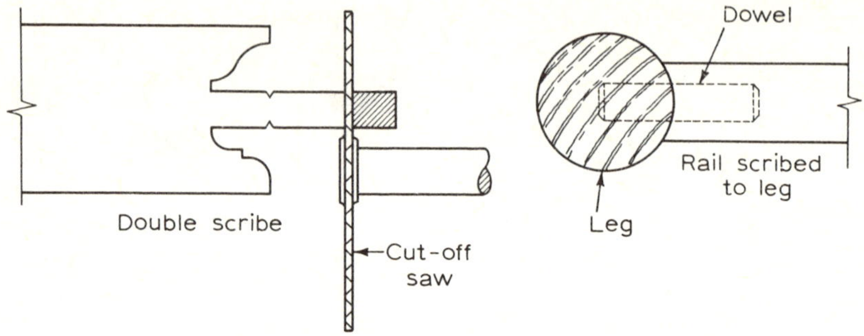

Figure 4.3 Machine scribing and cutting-off

the tenon is formed. If scribing heads are fitted (Figure 4.3) the table is rolled further forward until these cut in turn. Should a cut-off saw be fitted at the rear, still further progression causes the tenon to be cut to length. If the saw is at the front, excess material will be removed before tenoning takes place. The table is withdrawn and the rail is reversed for the second cut. This time a shoulder stop is used to produce the required length.

If work volume justifies the cost, a grooving head is sometimes fitted for the purpose of cutting twin tenons. Head arrangements usually follow predictable lines.

Chair makers favour machines with a front cut-off saw and two tenoning heads. Small joinery shops install machines with two tenoning heads and one scribing head (the lower one). Larger joinery shops will often have two tenoning heads, two scribing heads, a cut-off saw and if the machine is so designed, a grooving disc.

As the machine increases in size a swan-necked casting replaces the 'L'. From the front, or operation position, the general 'L' remains, but seen end-on the vertical 'L' is swan necked. Swan necking in machine terminology means that the main machine casting has top and bottom projections with a gap in between. With tenoners this means that the top head is fitted to the extremity of the top projection and the bottom head is fitted to the bottom projection, thus leaving a clear space in between. This space is utilised for cutting extra-long tenons that are longer than the length of the cutting heads, and for cutting trenches inward from the ends of wide boards for shelving, etc. Figure 4.4 shows the swan-neck principle and indicates its use. The 'L' machines are called *pillar* machines and are also shown in Figure 4.4.

The Machine Design

Most important of all (once initial design and accuracy have been built in) is the ease of setting. An expensive machine only earns its keep when it is fully utilised when work is available. Slow and clumsy setting methods, either inherent in the machine design or due to lack of skill of the operator, cost money.

The tenon heads must be adjustable easily and to reasonably fine limits. They must have relative movement to each other vertically and relative movement to each other horizontally. These aims can be achieved if one head has up and down movement together with horizontal movement. Two adjustment handles are all that will be necessary for setting. With vertical movement of one head and horizontal movement of one head, thickness of tenon and shoulder positioning is easily achieved. There must be one more movement, i.e. that to enable correct positioning of the tenon relative to the face of the work. This is achieved by one manufacturer with the inclusion of a handwheel and screw to raise and lower the rolling table. In this way, three control handles or wheels enable a full range of adjustments using the moving head for thickness, and the rise and fall table for position. On all tenoners, operators should think in terms of thickness and position.

The range of movements outlined, although ample for small machines, does limit the speed of setting where the work range is wide and short runs common.

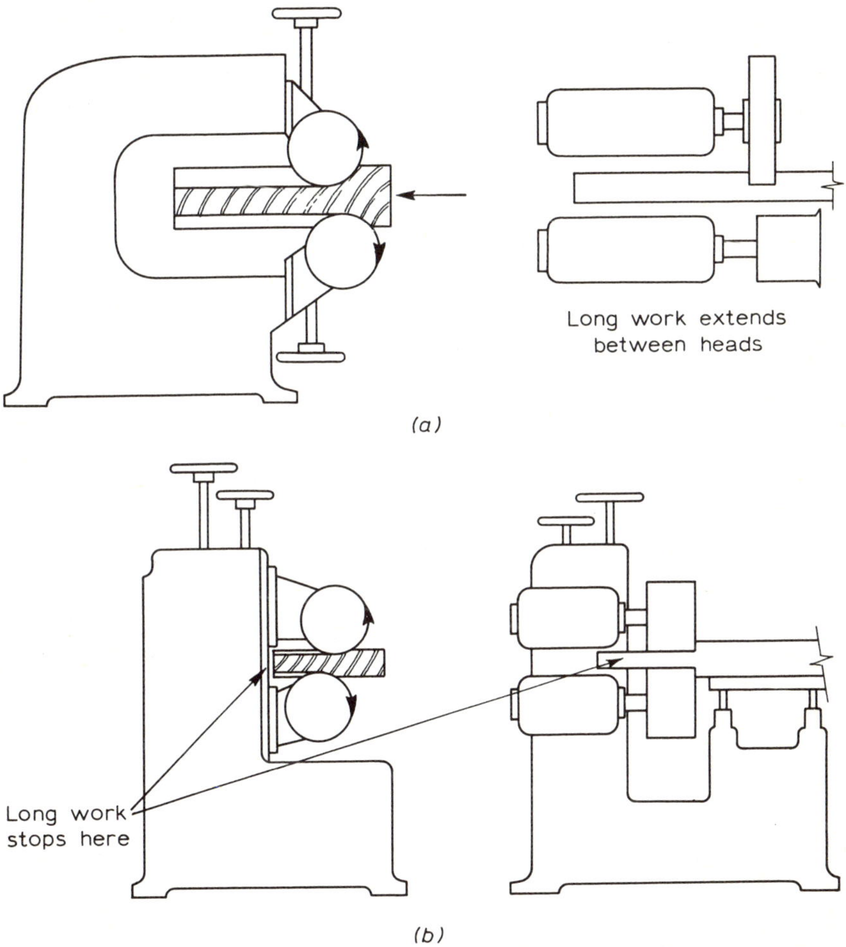

Figure 4.4 End outlines of (a) *swan-neck machine and* (b) *pillar types*

Improving on the design already outlined, two controls are added to the second head giving faster setting for thickness, but leaving the table rise and fall for position.

At this point the swan neck intrudes. Where the two heads are on separate castings with a full gap between them, no lead screws can cross this gap without obstructing the passage of wide and long boards. Inevitably, both top and bottom heads must have completely independent rise and fall. The table movement is not now necessary, nor a practicable proposition because of the weight of the heavier machine. Setting must be by skilled judgement, both heads being moved to achieve thickness and position. Alteration of thickness will mean alteration of position, and position alteration will require movement of both heads to maintain thickness. As additional equipment these large swan-necked tenoners have an across-the-heads locking bar. This drops down from the top head unit to the bottom head unit and locks them together in a manner which leaves them both controlled for position by one handwheel, and for thickness by the other. Thus two handwheels are used to set for thickness and position. Once thickness is correct, using one handwheel, positioning is attained by using the other which moves both heads without disturbing the fit. This device aids setting but destroys some of the swan neck advantage.

Reverting to the pillar type of machine (which, in effect, the swan neck becomes when the head lock bar is fitted) one method of making setting easier is to have the bottom head controlled for up-and-down movement by its own handwheel, which provides the thickness facility. The top head may not be moved without the bottom one. One handwheel controls this and also provides the position facility.

Thus tenon thickness is controlled by the bottom head handwheel, and tenon position is controlled by the handwheel that moves both blocks together. A variation of this arrangement is the design whereby the mutual linking of top and bottom heads for position can be broken by a mechanical clutch lever, enabling completely independent movement of both heads. Thickness is set using either or both wheels, and then the clutch is engaged enabling position to be corrected, using either wheel. Figure 4.5 shows some of these head-linking devices.

The horizontal tenoning heads are usually off-set, having centre lines staggered. This allows one head to commence cutting before the other, thus minimising the initial impact on the work. Each head on modern tenoners has its own built-in motor directly connected to the cutter-block shaft. With the perfection of the modern nylon/pigskin type of endless short-centre flat belts, some manufacturers are reverting to belt drives, which gives the opportunity to use standard motor casings and windings together with the opportunity to increase cutter-block speeds by adjusting pulley ratios*.

*The phrases *built-in drive*, *integral drive*, *direct drive* and others like *in-line drive* have crept into machine woodworking vocabulary without precise definition. They often tend to mean anything the user would like them to mean. The definitions of the terms used here are:

Built-in drive: where an electric motor is built into the machine for the purpose of driving one cutter head or gearbox.
Integral drive: as above.
Direct drive: where the motor is built round, or coupled to, an extension of the cutter-head shaft.
In-line drive: as above.
Indirect drive: where the cutter block is driven by belt.

It follows, then, that the first four phrases can mean the same thing, and that only the last should be used to indicate drive by a belt, and that it is possible to have a built-in indirect drive.

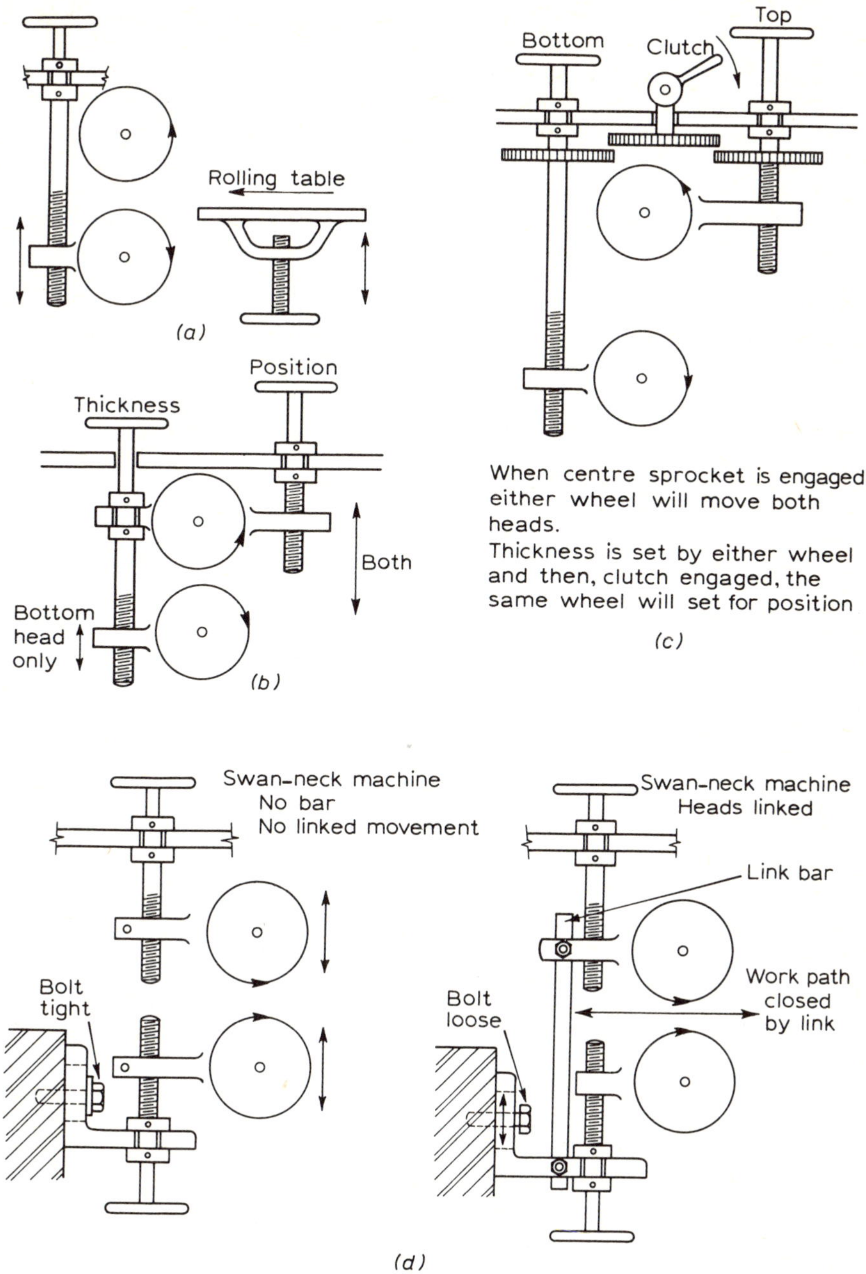

Figure 4.5 Tenon machine head linkage. (a) *One block and table height are the only adjustables on simple machines,* (b) *two-wheel control (one for thickness the other for position),* (c) *completely independent movement, both heads, or one-wheel control for both heads and* (d) *swan-necked head locking bar reproduces setting facilities of smaller machines*

The motors used for direct drive are usually slim-line and small diameter, and with the side towards the tenon face flattened as much as possible. This allows some overhang of long tenons on pillar machines. The motors are often completely sealed and lubricated for life.

Brakes are fitted to most horizontal heads and these are usually hand-applied friction pads operated by short levers. Danckaert of Belgium fit electrical brakes to all heads as an optional extra with the addition of a push-button spindle lock for cutter servicing. Figure 4.6 shows the off-set head plan, the slim-line motor arrangement, friction brakes, head locking and slide adjustment.

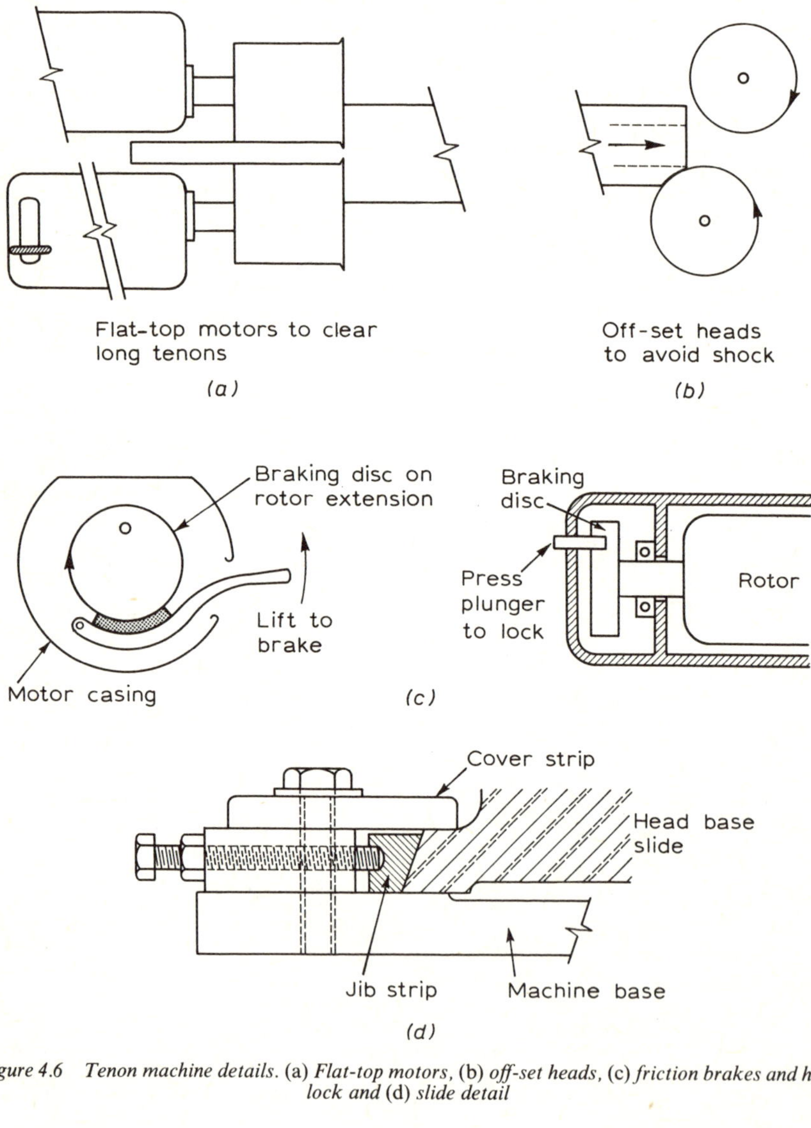

Figure 4.6 Tenon machine details. (a) *Flat-top motors,* (b) *off-set heads,* (c) *friction brakes and head lock and* (d) *slide detail*

Electrical Braking

Alternating current electric motors work on the principle that one rotating magnetic field (in the stator winding) reacts with the induced field in another winding (in the rotor). The first pulls round the second, causing rotation. The rotating field is a phenomenon of alternating current. If the current is cut-off, rotation ceases and the motor comes to a stop as it loses momentum. If a direct current, which does not cause a rotating field, is fed to the stator or outer part of the motor, it sets up a stationary magnetic field. This attracts the iron core of the rotor and pulls it to a stop. This form of braking is called *d.c. injection braking*. The machine control-box has a metal-plate rectifier which is brought into circuit when braking is required and which then feeds a direct current to the motor field windings.

Another form of dynamic braking (the name given to all forms of electrical braking) is known as *plugging*. This consists of reversing the rotation of the magnetic field in the stator. This opposes the rotation of the rotor and brings it to a stop and immediately starts it rotating the other way. Reversal of the rotation of a three-phase a.c. motor is achieved if two of the three wires feeding the motor are changed over at their connection. Thus plugging is arranged by having a switch that on actuation simply changes over two connections. When the motor reaches a standstill the operator cuts off the current. This system is not in great use as it requires close attention, for reversal of cutting heads could be dangerous.

The plugging principle is used, however, with slight modification. A three-phase motor will continue to run if one wire is broken after rotation has started. The motor will not generate much power and will begin to heat up. If the wire is disconnected before the starting switch is operated, the motor will not start; it will 'hum' and get hot. This is known as *single phasing* and it usually means that one fuse has broken.

Single phasing is used to brake motors. A motor running at full speed can be brought rapidly to a stop by the operation of a switch. This switch reverses two phases of the three and cuts the other one. This leaves a single-phase plugging field in the motor which opposes motor rotation but does not cause the motor to start reversing once it has reached a halt.

Electric motor principles are explained in greater detail in Chapter 7.

SCRIBING HEADS

Machines can be fitted with one, or two, scribing heads. These are fitted behind the main tenoning heads and are carried in slides built on the tenon-head castings, which means that movement of the tenon head moves the scribing heads at the same time. The scribing heads have their own horizontal and vertical adjusting screws and slide locks and it is only in the vertical that they move with the main head. The motors are built-in, being vertical and in-line with the scribing cutter blocks. Where only one head is fitted it is usually the bottom one.

CUT-OFF SAWS

Cut-off saws are positioned either in front of the machine or at the rear. The choice of position governs the choice of machine as manufacturers do not offer alternative positions. There is an element of sales promotion in the saw position as all manufacturers claim that the position they have chosen for the saw is best.

The arguments for front or rear positioning may be summarised as follows:

1. Saws in front cut off excess wood before tenoning, thus saving wear on tenoning cutters, and enabling the fastest possible feed through.
2. Saws in front can be used as cut-off saws without tenon heads being used at all.
3. Saws in front are more accessible.
4. Saws at the rear are safer as they are well protected by machine bulk. Saws are not always used and rear position enables free choice without removal from the machine and without laborious winding away.

Operators favour their own choice and this will depend on which machine they have been used to working. Technically the decision, 'fore or aft', depends on the nature of the general run of work; if there is enough work to use the saw all the time it is probably best at the front. Cut-off saws have horizontal adjustment only.

GENERAL FEATURES OF HEAD DESIGN

All heads have machined slides, some square, some 'V'-form. They all have adjustable jib strips for slide adjustment. All heads have slide locks and some have additional locks on vertical screws where vibration may shake a head downwards.

Between the main tenon heads there must always be a dead-stop to prevent blocks clashing. These are not infallible and need attention from time to time to ensure cutter safety. Machinists have a habit of screwing such dead-stops into their minimum position so as to fully utilise short cutters and to squeeze out the thinnest tenons for special jobs. When the new cutters are fitted, the dead-stop must be checked.

The large single enders have head-tilt adjustment to the top head. This is not for angular cutting, but to enable the top head to be made completely horizontal when used for deep-trench cutting. Normal design has the heads set slightly out of parallel to undercut shoulders. For square trenching this must be corrected. The device has a pin for location, and a locking nut. Guards move with the heads, but swing away for access. The Dominion Machinery Company have a form of guard on their *D.Z.* type of machine that swings away as the timber approaches the main tenon heads and swings back as the table is withdrawn. Most often, cutter guards are shaped to act as lead-in chutes for chip exhaust systems, and many machines have guards designed to fit into the standard equipment of a particular exhaust-system company.

Setting scales are often built in; sometimes on all heads for all movements and sometimes only for the main horizontal movement. No uniform system is followed for zero marking scales. Some have readings from zero inwards, and some from zero outwards. Some have central zero with readings in and out. The most usual for horizontal scales is to have the lower head scale zero at a point where the cutting head is at a safe distance from the edge of the rolling table. The upper scale has a central zero and the cutter head may be wound in for long and short shoulder tenons until it is actually cutting above the rolling table.

Setting technique varies, but in general operators use the scales as comparitors, assessing the amount of movement required from the setting positions of the last job completed to give setting positions for the next job along. For older machines without scales, operators or works engineers often fit their own scales. One device is to measure the pitch of the lead screw which gives the amount of movement for one complete turn, and then to paint divisions on the handwheel. If the pitch is 2·54 mm (0·10 in) then five marks round the handwheel will give controlled movements of 0·5 mm. It would seem reasonable to zero all vertical scales from the rolling table surface, but this is not always done. With rise and fall tables, scales can only be used as a guide for comparative movement.

The real problem of accurate machine calibration is the fact that cutters wear and that machine woodworking cutters are, in the main, not part of the block on which they run. Cutter setting inaccuracy renders the most precise scale inaccurate.

Setting methods will be outlined later.

THE ROLLING TABLE

The job of the rolling table is to allow the accurate positioning of the workpiece, give adequate work holding, and to allow accurate forward feed through the cutting cycle.

Track design is important. To give good work, the table must roll easily and accurately, and have its movement immune to deflection by chips. Racking should be negligible; a good table should roll if pushed from either corner. Generally, one trackway is used to govern straight-line control with the other of the pair simply supporting the weight. Two track designs are shown in Figure 4.7.

On the table is a back fence against which the work is held when being cut. The fence will pivot forward for angular shoulders and have accurate stud location for return to square. The fence carries a stop bar which is fitted with swing-over shoulder stops. Holes through the fence allow a semi-permanent wooden-back fence to be screwed on. A softwood-back fence is usually fitted by pinning after the machine has been correctly set and tested. This saves operator time, for the hardwood strip is not often replaced.

In the bed of the rolling table there is usually a pop-up spring-loaded shoulder stop, carried in a bar casting that can be moved along on slides. Across the table, front to back, and at the cutter-block end, there is a clamp bridge that carries the work clamp. This clamp can be full length across the table and operated by one hand-control lever, or there can be two small clamps that may be adjusted to suit job width. Most clamps are operated by lever and eccentric cam with a flat on the cam just beyond the maximum pressure point. This causes the clamp to snap-lock as the hand lever brings the pressure pad down to the work face. Pressure pads are usually faced with a resilient pad. One fault of the single-

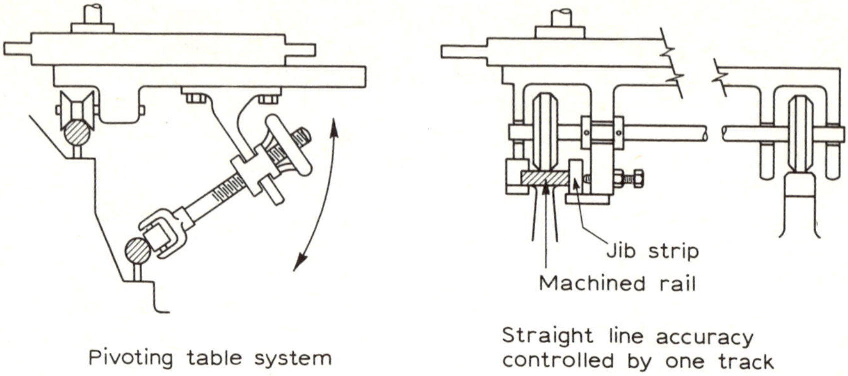

Figure 4.7 *Rolling table track designs. These vary, but usually only one track controls straight line accuracy*

pressure bar clamp is that it is often made too long, with its back end so close to the metal fence that it is difficult to screw-on a false back fence.

A first-end work stop is usually positioned on a projecting arm, level with the work fence when the table is fully withdrawn. The minimum tenon length is immediately ensured by pushing the first end of the work against this stop. On turning the rail round for second end tenoning, this stop is swung back clear. The use of these table fitments will be discussed under *Setting Methods*.

An additional movement to the work table is added by the manufacturers of the smaller single ended tenon. The rolling table will tilt, and this is limited to a few degrees up but provided to a full 45° down. Tilting rolling tables may only be fitted to machines where the table tracks are vertically above each other.

Alternative design

Several machines for single ended tenoning have only two heads. These have their centre lines vertical and the cutting head of the top one is at the bottom of its motor shaft (or drive shaft where short centre belts are used) and the bottom head is at the top of its shaft. The cut tenon runs between the two heads, which can have disc-type blocks or conventional square blocks. Scribing cutters for shaped shoulders are carried on the same head that cuts the tenons. Cutting end-on to the grain fibres is heavier on the motor and cutters than conventional cutting is and motors are usually a little more powerful. It is this type of machine that is usually fitted with the rolling table runners in-line above each other, and has the capacity to tilt.

Round ended tenoners

The Rye range includes single enders and double enders, and all of the machines are primarily intended for furniture manufacturers. The Rye principle, whether single or double end, is that of a moving cutter block travelling in a path that will produce a tenon of the width and thickness required, and with tenon sides rounded to the radius of half the thickness of the tenon. Such a tenon will easily fit into a rounded-end mortise cut on a router or jigger mortiser.

The path of the cutter block is controlled by a cast track that forms a groove into which a cam fits. The track path has two short straights connected by two semi-circular ends. The track is divided along a horizontal centre line, and set screw adjustment moves the two halves apart to increase the thickness of the tenon. There is a slight discrepancy of the round at the tenon side caused because the two semi-circular ends do not alter their radius as the two cam-track halves are opened to give thicker tenons. This is so small that it is of no consequence. The tenon head is large, approximately 100 mm in diameter, and compared to the radius of the tenon side it is so large that such discrepancy of radius is lost.

To increase the width of the tenon, spacing pieces are inserted half way along each straight portion of the cam track. With no inserts, and the track closed up, wood pins can be cut on the ends of rails instead of tenons.

The cutter block, which is circular, carries two straight cutters for the tenon, two shoulder scribers and two spurs, one of which trims the tenon to length, and the other of which chamfers the edge of the tenon end to facilitate assembly.

THE PROCESS

All components of the work cycle are automatic, with compressed air controlling sequencing and two electric motors providing drive power. The tenon cutting head is driven by 'V'-belts from its own motor. The other motor drives the cam (and cutting head) along the path of the cam track.

Work is loaded against a fence, which is positioned on the work table to give the correct position of tenon related to rail width. The work table has rise and fall adjustment to facilitate position setting relative to the rail face. Both table and fence have the pivoting ability essential for compound-angle chair work.

When work is positioned, depression of a foot pedal (multi-path air valve) causes air to be fed to the table cylinder, which moves the table in towards the cutting head. As the table moves, a trip bar opens an air valve that feeds air to the work clamp cylinder. The table continues to move forward bringing the rail end into the cutting orbit of the cutter block. The block is positioned dead centre to the required tenon with the cam in a similar position in the cam track. The table moves forward until the cutter block has cut the required depth into the rail. When this point is reached the table stops, and at the same time operates another air valve that controls a small air cylinder in the clutch mechanism.

The clutch has to be held out by the air in the cylinder and as the control valve, operated by the table edge, releases the air, the clutch engages and the cutter-block head movement mechanism starts to operate. The clutch consists of a hard rubber disc that is slightly smaller in diameter than the recess in which it runs. As the air cylinder moves, it allows the rubber disc to contact the rotating pulley in which it is situated. Drive is transmitted from the driven pulley through the rubber disc to a gearbox that starts the cam moving in its track.

The cam follows the track, and the cutter block moves up, round an arc of 90°, onwards across the top face of the tenon, down and round an arc of 180°, and then back across the tenon bottom face. A 90° upward arc then returns it to its start position. When the cutting head has ended its stroke, a trip valve is operated from a cam on a rotating disc at the gearbox, and an air signal goes to the foot valve, which then reverses. Air is then fed to the table cylinder. The table withdraws, the air clamp releases and the rail is ready to be taken out, reversed and repositioned with the shoulder against a shoulder stop. Time cycle is from 4 to 7 s.

The limitations of this sort of machine are due to the physical limitations of the cam track which must be related to the cutter block size. Any great increase in tenon thickness necessitates a smaller cutter block. The double-ended version is similar, but has a separate linkage to drive each head so that each end of the rail can have a different tenon. To achieve angular tenons the machine main castings are moved apart and two workpieces are inserted together, each at its own appropriate angle.

Later versions of the Rye have the cam track groove positioned with its straight sides vertical. The operator stands end-on to the rails being cut, which are on edge. One rail is inserted against a fence, which is one of a pair. When the foot valve is operated the table moves across the machine face into the cutter head orbit. The block travels round, up, round and down, completing one tenon. Meanwhile the operator has placed a second rail in position against the second of the pair of fences. When the first tenon has been cut the table moves back across the machine to the unload position. At the same time, the second fence has aligned the second rail into the cutting position. The work cycle is very fast, loading being done whilst the machine is cutting. Output can be as high as 700 tenons per hour. The cam track design is radically different to the first machine described and a variable cam is used to control tenon thickness. Explanatory diagrams of the Rye tenoners are given in Figure 4.8.

Double ended tenoners

'Double enders', as they are known, are complex. No two are the same, for such is the range of equipment available that each manufacturer offers the basic machine castings and slides, etc. and the purchaser specifies the equipment to be fitted. They all conform to a general picture: each has two main machine heads, one of which is mounted on slides, to move closer to, or further away from, the other. Between these machine heads there are two chain beams, fitted from front to back, and on which run the feeding chains. Above the feeding chains are track pressures to hold workpieces steady whilst they are cut. The pressure can be in the form of caterpillar chains made up of hard rubber blocks, or they can be endless bands of stainless steel backed by sprung roller pressure. A later development, fitted to machines in the Wadkin range, has a synthetic block chain that runs over a spring steel strip backed by foam rubber for the provision of adaptability for slight rail width variation. These top pressure chains may be free running, either driven by friction from the work top surface or power driven, with the drive synchronised with the lower feed chain. This ensures continuous feed where backward-facing dogs are used. These are pushing pieces fitted to the feeding chains and it is against these that the work is

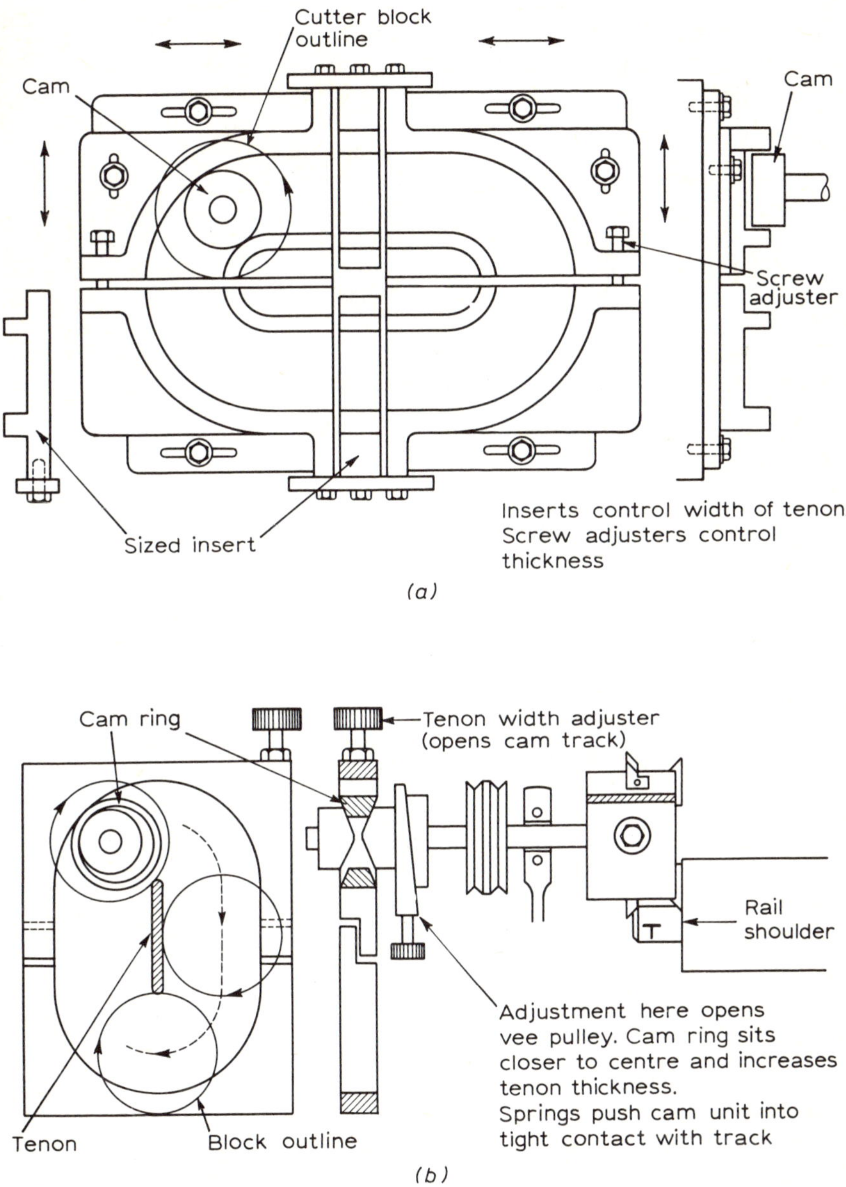

Figure 4.8 The Rye rounded end tenoner. (a) *Control system used on earlier single enders and on larger double enders, and* (b) T3 *cam track system used on the later single enders*

positioned. They are spaced out at intervals on the feed chains. To make it possible to feed panels wider than the normal dog spacings, alternate dogs may be 'disappearing' and these fall below chain level when work is placed on top of them as is the position with wider rails and panels.

When very wide (or long) panels are fed, chain dog spacing is not wide enough to accept them and neither are the front extensions of the chain beams. In this situation dogs are 'backward' facing, with their location faces opposing the feed direction. The chain links have friction pad inserts. These, with the downward pressure from the pressure chains, are sufficient to grip panels for forward feeding. One chain of the pair has adjustment for position relative to the other so that dog pairs can be squared across. One range of dogs is also adjustable so that single dog pairs can be adjusted.

Each main machine head carries the cutting head units. As with single enders, these consist of cut-off saws, tenoning heads and scribing heads.

This is the basic machine. The main differences at this stage between single and double enders are that on the double enders there is no linked movement between the heads for vertical adjustment; the cutting heads usually pivot 45° up and 45° down, and there are usually much more accurate setting scales, sometimes of the micrometer type around each lead screw. Often too, the horizontal slides have vernier scales. On some of the smaller machines the tenoning heads may pivot through 90° so that they can cut as normal horizontal heads, or they can cut edge on (with the main axis vertical) and have scribing cutters fitted to scribe at the same time as tenons are cut.

Given the basic machine, setting-up and working is almost the same as with single enders, with the difference that work is placed on a through-feeding chain instead of a push-pull rolling table. Cut-off saws are at the front, and a guide fence is used to position work ends.

Feed-chain speeds are controlled by a gearbox, two-speed motor or stepped cone pulleys. The usual range of feed speeds is from 3 to 25 m/min. The distance (pitch) between dogs may be as low as 200 mm. Given that only alternate dogs are used and rails are fed in pairs, output at 6 m/mm would be 30 complete rails per minute. Combinations of dog pitch and chain speeds together with the number of pieces reasonable for an operator to pick up and position, give the machine an enormous potential if fully utilised. It is no advantage to have close dogs and high chain speed if the operator can only catch every third or fourth dog.

Such is the basic double ended tenoner. A reasonable all-round machine man can soon master its control and operation.

THE DOUBLE ENDER AND PROFILER

Equipment can be added to the basic machine that will make it possible to perform work normally done by other machines. Examples of work functions possible are:

Panel sizing | Rail haunching

Panel edge moulding	Housings: stopped, through and dove-tailed
Panel edge shaping	Drilling: face or end-on
Panel tongueing and grooving	Corner locking
Panel tongue relishing (haunching)	Mitre sawing
Panel end stopped grooving	Interlock corner joints
Panel lip trimming	

This range of work is made possible by the inclusion of cutting heads carried by the chain beams, overhead cross beams, or on the main vertical slides. A short survey is all that is possible.

Hog saws. Saw and cutter combinations that reduce all off-cut waste to the size capability of the extraction systems.

Scoring saws. Fine-gauge shallow-cutting saws that penetrate the surface veneer in-line with follow-up tenon heads to ensure clean shoulder cuts. Can be 'jump-up' air controlled.

Relishing units. Small air-controlled cutter blocks designed to pop-out or jump-up at planned positions to cut haunches or reduce ends of tongues. They can travel with the chain for short distances as they cut, to give square cuts.

Gaining heads. Carried on a crossing overhead beam these heads carry end-cutting (router-type) bits, or side-cutting disc cutters. They cut housings, and can jump to cut stopped housings. They can be adapted for drill work. When drilling into a moving panel they move back with the feed and jump forward as they withdraw from the hole bored.

Profile units. Positioned on chain beams at the rear of the machine. These move in or out, and are controlled by a rotating cam or straight-line slide cam. They are synchronised to chain feed and dog spacing in such a way that controlled contours are possible.

Transfer equipment. Two double enders can be linked at right angles or in-line to enable four sides of a panel to be cut at one pass. The unit accepts panels and either turns them through 90° or feeds them directly sideways to the next machine.

Stackers. Units that receive cut panels and stack them.

Hoppers and feeders. Units that hold a stack of workpieces and dispense them one at a time to the machine.

Movements of all of these units are controlled by a combination of applied pneumatics, electricity and mechanics.

A rotating cam drum, with cam projections, operates a bank of micro switches that in turn operate solenoids that control air-flow valves that govern air flow to working cylinders that have mechanical links to the part that is to move. Thus rotation of the drum, which is gearbox linked to the main feed-chain drive, controls and synchronises all movements. Figure 4.9 shows a schematic layout of two basic control systems.

Other electric contact switches or air trip valves can be positioned along the chain beams and these may initiate movement of cutting heads when touched by moving workpieces. Transfer units and stackers are often so controlled, but can have in addition photoelectric cells actuated by the leading edge of the workpiece.

There is no reason why such complex machines should not be controlled by punched tape scanners. There are already machines being marketed that reset

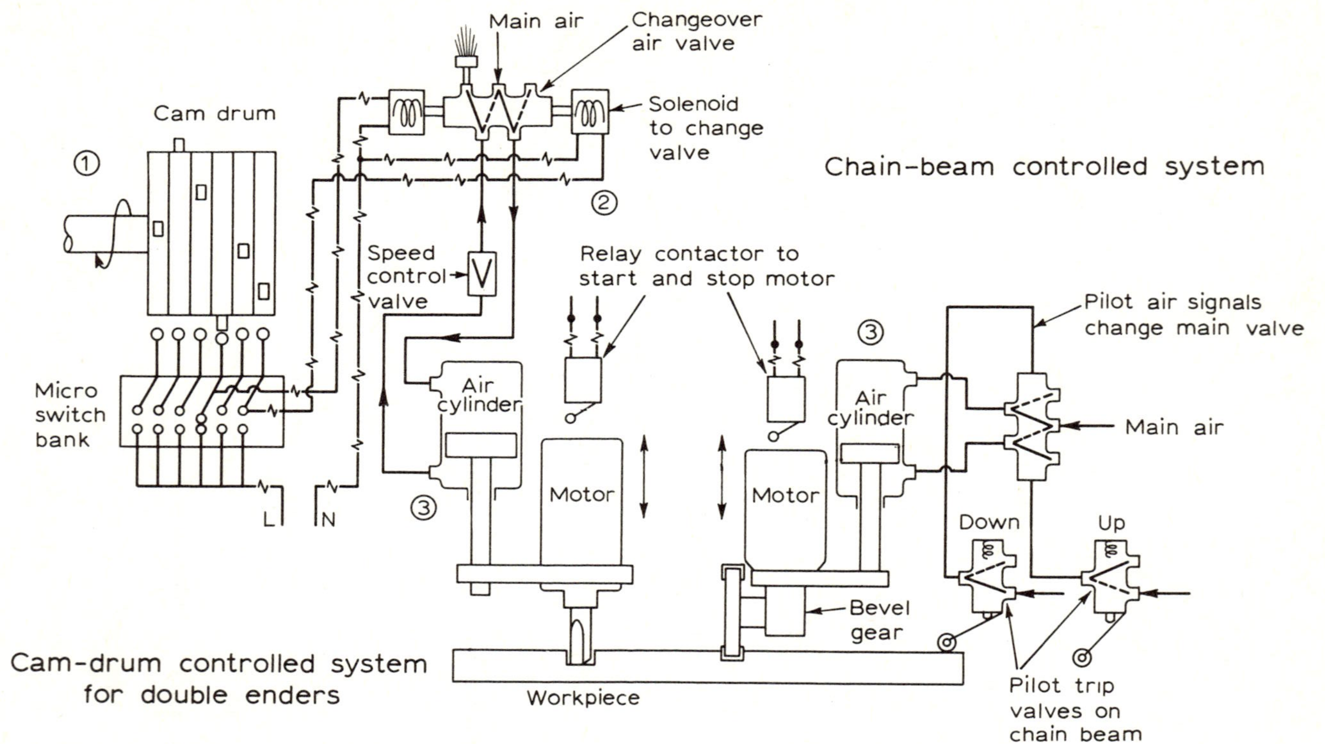

Figure 4.9 Basic control system for 'jump' head control on double enders.

1. Cam drum and micro switches
2. Solenoid-controlled air valve
3. Air cylinder actuators

Additional control for other units may be from air trip valves on the chain beams themselves, job to job, as instructed by tape scanners. It must be remembered that machines cost money to buy, run, equip, maintain, light, heat and replace. Non-productive time earns no revenue to pay for any of these.

Machine setting

SINGLE ENDED TENONERS

Setting to pre-cut mortises is the best and safest way, for even stock setting pieces deteriorate. The setting system outlined below is tabulated, but experienced operators work all movements towards the required end, and do not necessarily move only one head at a time. Examination questions relating to processes are often best answered in the following tabular form:

1. Use the mortising tool to mark the rail end.
2. Set the tenon table clamp.
3. Position the setting rail close to the cutter block ends.
4. Swing the blocks by hand and adjust until the main tenoning cutters line-up by sight with the track made by the mortising tool.
5. Cut the test tenon on a piece narrow enough to go into the mortise.
6. Try the tenon fit and check the shoulders.
7. Adjust for fit and shoulder squareness.
8. Reset, readjust and then set for position.
9. Reset, then set for scribe, if any.
10. Nail the false back fence to the semi-permanent fence. Project this false fence under the cutter block.
11. Cut the tenon on the false fence.
12. Trim the end of the false fence to the minimum tenon length, and set the tenon length stop to this.
13. Set the first test rail shoulder line to the shoulder of the false fence. Cut the tenon.
14. Reverse end to end, keeping the face side down and put the shoulder line to the fence shoulder.
15. Cut the tenon. Check the shoulder length and set the swing-over shoulder stop.
16. Commence work.
17. Send the accurate test piece (if a scribed job) on to the spindle machinist.

When working, keep all guards closed and all slide locks tight. Keep the left hand on the clamp lever and the right hand on the back end of the work.

Tenoning cutter blocks

Traditionally, horizontal tenoning blocks have shear cutting cutters. This means that the cutter seating is arranged obliquely to the block axis. The theory of this is that the cutter edge is introduced progressively to the cut thus creating less shock and vibration, and giving a situation where the wood fibres separate more easily. The resulting surface of the tenon should be smooth, and without lifted grain.

Shear cutting, in general, is a proved method of improving cutting performance in difficult conditions. Where tenon blocks are concerned the idea looks attractive, but in fact, it is largely based on a fallacy. Mr. Peter Mirish of the Liverpool College of Building undertook extensive investigation of tenon cutting blocks in 1959 and concluded that such was the wide variation in tenon block shear angle that its advantage was not proved. Why else the great variation in angles? As a result of his work, Mirish came to believe that what was necessary was a low and constant angle of approach by the cutter to the wood, rather than the oblique approach through an ever changing angle of shear cutting cutters.

As a result of this investigation, Mirish co-operated with others in the design of a rectangular cutter block that gave the cutter a 50° cutting angle throughout its length. This cutter block has been offered on the machines of Thomas Robinson of Rochdale since it was first designed, and tenons cut with it compare very well with those cut on traditional blocks.

To revert to the standard block with its oblique seating: it has two flat faces for the main tenoning cutters, and these are out of parallel to each other and to the block axis (Figure 4.10). This angled seating creates a changing cutting angle from one end of the tool to the other. An oblique cut through a cylinder produces an ellipse and to cut a straight tenon with a cutter set obliquely to its axis of rotation requires that the cutting edge generates a cylinder as it rotates. For this to happen the cutter edge must be shaped to an elliptical curve.

The curvature of the cutting edge will vary as the angle of the cutter seat to its axis varies. A short block with high angled seating would need a cutter with greater curve than would a longer block with less angle. Consequently, all horizontal heads with shear cutting blocks require templets for the cutter curvature, and great care in setting. Chips removed by these blocks are usually as long as the tenon being cut and tend to block the exhaust system. To overcome this and to reduce vibration in cutting, large machines have split blocks with four cutters, which reduces chip length. With longer, single blocks, operators sometimes introduce a 'nick' or 'gap' in each cutter. This creates a chip with a line or raised bead across it. The following cutter, with no gap, removes this bead and weakens the chip so that it comes away from the workpiece in two pieces.

On the shoulder ends of the blocks are two flanges, each carrying a small toothed spur cutter. Loose terminology has caused this to be styled a *scribing* cutter. As this cutter inscribes or incises a line across the work face in advance of the main cutter, the name is reasonable, but as trade terminology also refers to the act of marrying one moulding to another as *scribing*, and because this is done, in tenoning, with special blocks known as *scribing heads*, the name *scribers* should not be given to spur cutters. Perhaps a better name for these is *shoulder* cutters. These shoulder cutters are set slightly in advance radially of the main cutters, and slightly in advance longitudinally. Shoulder cutter faces are inclined in towards the block centre so that only the extreme tips cut. This setting is shown in Figure 4.10, together with a geometrical method of determination of the elliptical cutter profile.

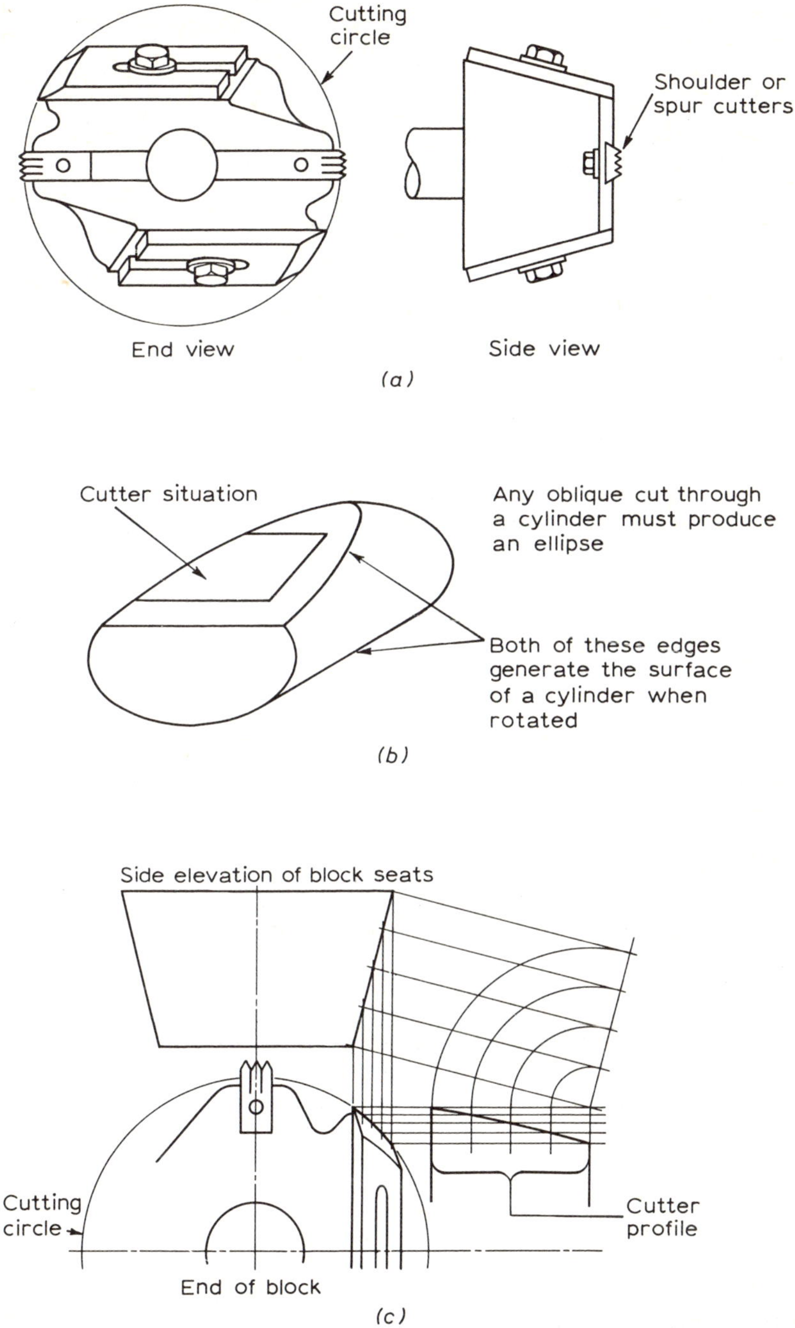

Figure 4.10 Shear-cutting tenon blocks. (a) *Tapered cutter block,* (b) *reason for elliptical cutting edge and* (c) *geometrical development of this shape*

SCRIBING HEADS

Blocks used for scribing are most often square. The trend is to change this and to fit circular disc cutters that are more easily set, and to use less costly cutters. Generally, scribing blocks are set with their top faces just clear of the tenon cheek that they are to scribe. The cutters are then tilted so that their leading points just clear the tenon. It is bad scribing practice to allow the scribing cutters to leave marks on the tenon face; this spoils the preset thickness of the tenon and leads to undue cutter wear.

Tool maintenance

Spur cutters are ground, on their backs, to a sharp edge. Their edges are curved and must conform to an arc of the cutting circle. If the block is rotated by hand against a setting piece of wood this shape is easily checked.

The main tenoning cutters are ground free hand to a prepared templet usually provided by the tenoning machine supplier. They need a fairly wide clearance angle to allow the edge to be thin and sharp, and to allow for the changing cutting angle end to end of the block. After grinding they are slipstoned.

Scribing head cutters are ground free hand with an allowance made for loss of profile due to angle of approach to the wood. The principles related to cutter profile change due to changing cutting angles is discussed in Chapter 6.

Faults in cutter setting

Tenons must be flat on each face, parallel in thickness, and parallel to the rail face. If the tenon face is round from shoulder to end, the tenoning cutters have insufficient curvature. If the tenon face is hollow, the cutter has too much curvature. Tenons tapered or out of parallel are due to bad cutter setting and this must be rechecked.

Rough or burnt shoulders indicate that the spur is not set correctly. It can have insufficient projection, or the main cutter can be set too far along the block so that it cuts beyond the spur. Burning usually means that the spur is projecting too far, or has an edge that is too thick. Careless slipstoning on the spur face can easily cause burning because the extreme leading edge has been dubbed over.

Motor power

Most horizontal tenoning heads have motors of approximately 1500 W but this tends to vary from machine to machine, with some manufacturers offering motors of 3200 W output. The usual speed is 3000 rev/min less 'slip' (see Chapter 7 for 'slip').

Calculations

The only calculations associated with tenoning are those related to the output of double enders. Hourly output is equal to the chain feed in metres per minute divided by driving dog pitch and multiplied by rails fed per dog and by 60 and 1000.

Example 4.1

How long would it take to produce 15 000 rails if the chain was set at 15 m/min and the operator was putting five rails against every fourth dog? Assume uninterrupted working, and dogs spaced at 200 mm.

$$\text{Time taken (h)} = \frac{\text{No. of rails} \times \text{Dog spacing (mm)}}{\text{Chain speed (m/min)} \times 1000 \times \text{No. of rails per dog} \times 60}$$

$$= \frac{15\,000 \times 800}{15 \times 1000 \times 5 \times 60} \qquad \text{(dogs at 200 mm)}$$

$$= 2{\cdot}66 \text{ h}$$

$$= 2 \text{ h } 40 \text{ min (approx.)}$$

5 Narrow Bandsawing

Definition

Narrow bandsawing machines are machines that use 'narrow' bandsaws for the purpose of cutting curved and irregularly shaped outlines from mainly flat boards and sheet materials. Narrow here means from 6 mm up to 38 mm wide. It is possible to obtain and use saws as narrow as 4 mm, but these are exceptions to the normal. The upper width for normal work finishes at about 25 mm, and again, the wider saws are for exceptional work.

The name *Wide Bandsaws* met with in the woodworking industry relates to those saws, normally over 75 mm wide and up to 360 mm, used for the purpose of converting round logs into boards on band-milling machines, or for re-sawing large boards into smaller sectional ones on machines known as *band re-saws*.

Narrow bandsawing

The machine principle is that an endless steel band, toothed at one edge, is run around two large wheels, tyred with hard rubber, and tensioned apart to give strain to the saw. The lower wheel is the driver, pulling the saw downwards through the wood. This lower wheel is mounted directly on the shaft of the driving motor which is usually wound to give the correct speed through the rim of the wheel. The main casting of the machine is swan necked, giving overhang to the top wheel, and this enables a large work table to be positioned horizontally between the wheels. The saw runs through a slot in the work table. The top wheel is adjustable up or down to accommodate varying length saws and to enable saws to be given the correct strain. The top wheel is also arranged to pivot forward or backward to facilitate saw tracking.

Other essential features are: the saw guides above and below the table, thrust wheels above and below the table, a wheel brush for the lower wheel, wheel brakes, and of course, guards.

The machine size is given by the diameter of the saw driving wheel. The most common machines are those which have wheels of 762 mm diameter. Larger machines for heavier work have wheels of 915 mm. Occasionally industrial

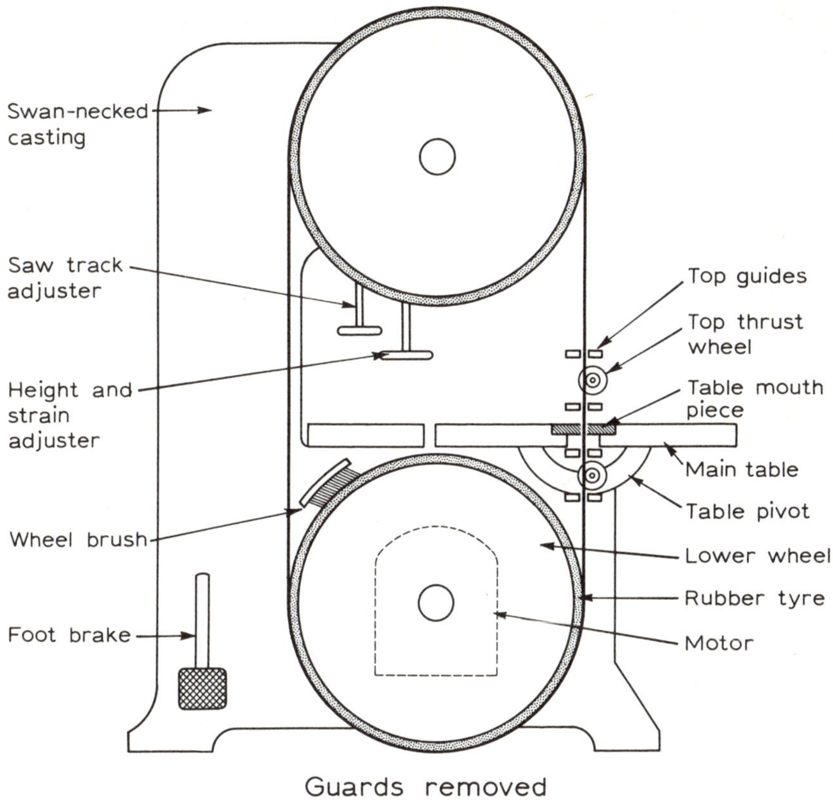

Figure 5.1 Main features of a narrow bandsawing machine

workshops have machines with wheels of 610 mm, 560 mm and 455 mm diameter. Figure 5.1 shows the various parts of the narrow bandsawing machine.

The saw

Narrow bandsaws are made of carbon steel strip with varying amounts of nickel and other metals alloyed to give greater wear resistance. The saws are joined to form an endless band either before purchase, or, in large factories, in the tool maintenance room. The method most often used is that of electrical butt welding, sometimes known as *flash* welding. An alternative method, but one which requires a higher degree of craft skill, is that of lap brazing, again by electricity.

With the butt weld method, the saw is cut to length with the set of the teeth matched. Setting of teeth is, as with circular saws, the alternate sideways springing of teeth to give side clearance in cut. If the set is not matched and the saw is later reset automatically, the machine doing the setting will, when it meets the two teeth set to the same side, start to push all teeth back to their opposite sides. Having cut the saw to length using small machine shears, it is positioned in the jaws of the welding jig. The saw ends are positioned to touch, and the jaws holding them are held apart by this touching of the saw ends. The jaws are set to snap in towards each other as the weld is performed, and the amount by which they move is called the *jolting distance*, and a chart attached to the jig gives the correct amount of jolting distance for each width of saw. The saw butt joint is now the point of greatest resistance in a low-voltage high-amperage electrical circuit. The voltage varies with machines, but is usually about 12 V, and can be varied by switching to suit differing widths of saw.

As soon as the circuit is closed by switching, the current flows, and, meeting high resistance at the butt, it causes a sudden and intense rise in temperature. This melts the saw steel at the extreme ends and the sprung jaws push them together, breaking the circuit at the same time. The metal of the saw cools rapidly and the saw is joined. However, the saw being of carbon steel, the sudden increase and decrease in temperature causes hardening and stressing to take place in the vicinity of the joints. To overcome this potential source of trouble the machine jaws are positioned wider apart and are free to move in or out with the saw which they clamp. A restricted voltage is then passed through the joint and this raises the temperature sufficiently to anneal the joint area and to soften any hard spots. The aim is to have the joint area slightly softer and therefore tougher, than the rest of the saw.

Both sides of the saw will now have slight beads of metal pushed up by the closing jaws during welding. These are filed off and the saw is checked for straightness. Any roundness of the saw back or tooth line is corrected by spread hammer blows along the opposite edge. After sharpening and setting the saw is ready for use.

With the brazing method, the saw is cut to length with the set matched, and the ends to be joined are filed to make a scarf lap. One tooth lap is sufficient and is easier to file than the two teeth lap that used to be recommended. The saw is placed in the jaws of a jig similar to the welder, but with the addition of a pair of clamping jaws, manually controlled, that can squeeze the lap area. The teeth are matched for position and the whole joint area is coated with a soldering flux, usually borax paste. A small piece of hard solder, cut from a strip, is placed between the ends of the saw, and it should be sufficiently large to project all round. This hard solder is specially prepared for bandsaw brazing and is about 75 μm thick (0·003 in).

The joint is again the point of highest resistance in a low-voltage high-amperage circuit. The circuit is closed by switching and the joint heats. Solder flows and the operator closes the joint clamp and switches off the current. After a few seconds the clamp is released and the joint is inspected. If the joint appears good the saw is placed on a filing block and excessive solder is filed away. The saw is then returned to the brazing jig to be annealed, which means that current is again allowed to flow through the joint, but only sufficiently to bring the bright, filed, metal up to a dull red. This reheating anneals and softens hard spots.

The joint is again filed, this time very carefully to bring its thickness to that of the rest of the saw. The back edge is filed straight, and the teeth are filed sharp. Subject to setting and sharpening, the saw is now ready for use. Figure 5.2 shows the two methods of joining narrow bandsaws, and Figure 5.3 gives details of saw teeth.

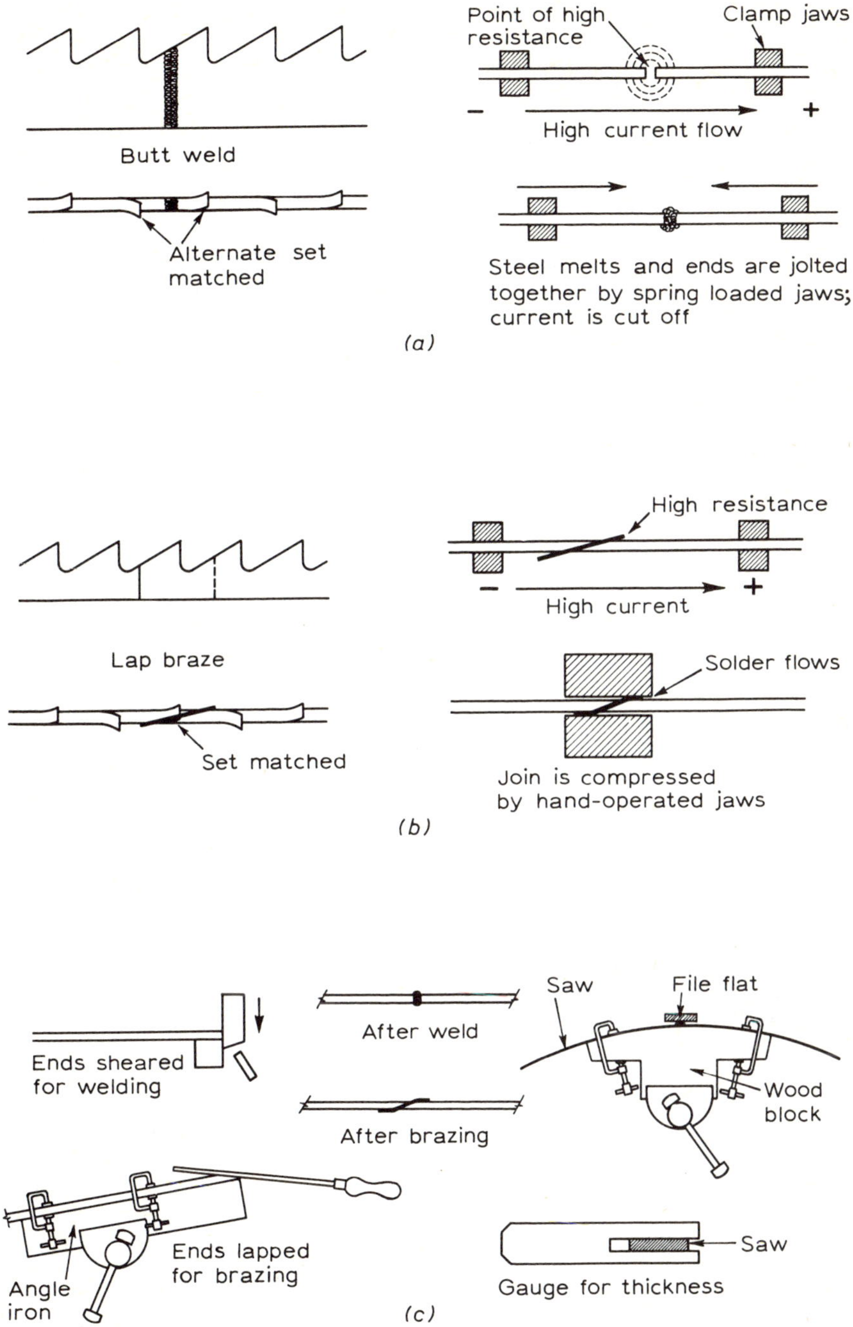

Figure 5.2 Two methods of bandsaw joining: welding and brazing. (a) *Electrical butt welding,* (b) *electrical brazing and* (c) *cleaning methods*

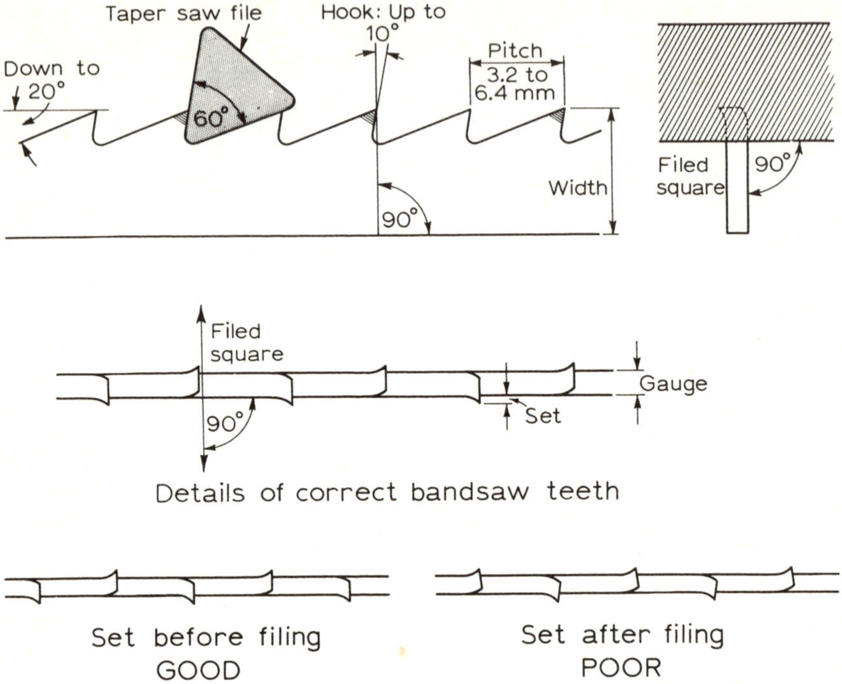

Figure 5.3 Bandsaw teeth. These are triangular, as shown; as the pitch increases, the tooth gullet must get deeper or the angle of the hook must be increased; too much hook leads to snatch

SHARPENING

Saws are filed to sharpness by hand or machine, although the trend is towards machine grinding. Both for hand and machine use, the files are of a type known as *taper saw files: second cut, single*. They are triangular, with rounded corners and have only one cut of teeth, of grade two. Engineers' files usually have two cuts of teeth at alternate angles, and, if triangular, have sharp corners. Taper saw files have rounded corners so that the tooth gullet is left round, and has less tendency to crack.

With machine filing, the saw is supported in jaws and the machine is set for four factors: the file angle must suit the hook angle of the saw; the feed pawl must feed each tooth only the correct pitch distance; the feed pawl must leave the tooth face in line with the file face; and then the back support must be set to give the right thrust to the saw to enable the file to make only a light cut. The machine will now file all teeth equally and, if a mark is made at the commencing tooth, the operator in charge will be able to check the saw's progress. For machine grinding the process is similar but a small grinding wheel is incorporated instead of a file.

For hand filing, which is faster and generally produces a sharper saw, the saw is held in a long horizontal vice, and each tooth is given one stroke of the file. The saw is progressively moved along the vice. Each tooth is filed in exactly the same way, as bandsaw teeth have no front or back bevels, both fronts and backs being filed square across.

SETTING

Bandsaws are set to give side clearance. They are spring set, alternate teeth being sprung to opposite sides. Setting can be done by hand (strictly for experts) using a light hammer and special anvil, or by machine. The filing machine can be adapted for setting, or, as is more usual, an additional setting machine is arranged to work in conjunction with the filer, and so complete the job in one pass. The amount of set required will vary according to the nature of the material being cut and to the width and general condition of the saw. Softwoods require more set than hardwoods, narrow saws require less set than wide ones, and old, slightly twisted and buckled saws require most set of all, and are useful only for rough cutting jobs. Set varies from 300 μm to 390 μm* (0·012–0·015 in).

It is usual to sharpen before setting, and this is acceptable, but theoretically setting should be done first because this method produces the chisel-edge teeth that a bandsaw requires. If setting is done after filing, the square face gets sprung slightly out of square.

FOLDING

For storage, narrow bandsaws are folded into three loops. The saw is held in two hands at about chest level, with the teeth pointing away from the body. With the fingers and thumbs, the teeth are twisted outwards and the top loop is gently persuaded out, and down, to meet the bottom loop. The saw is now part folded, with two half loops, teeth face to face, at ground level, and two half loops at waist height, teeth facing outwards. The hands are brought together and the half loops passed one over the other. As these half loops are pulled away from each other the teeth are now facing; the two bottom half loops begin to move towards each other. With gentle persuasion they both rise to lie horizontal and the hand-held half loops are brought down on top of them, thus giving a saw that is folded into three complete loops. Other hand movements can produce the same effect, and it is possible to fold a saw using only one hand.

To unfold, careful selection of holding points is necessary, as a saw that is simply shaken will spring, and if wide, spring violently, to the danger of anyone close to it.

The bandsawing machine

The top wheel is very important to success in sawing. It has four movements: (*a*) it rotates when driven by the saw acting as a belt and transmitting the drive

*The Greek letter 'mu', printed and written as μ, indicates in S.I. a multiplier of one millionth, and is spoken of as *micro*, associated with its appropriate unit. The amount of set given here for narrow bandsaws should be spoken of as 300 micro-metres. However, there is no immediate likelihood, nor any distant probability, that any shop floor workers will have suitable equipment, ability or instruction, to measure such minute units. It seems that micro-metres will be of no practical use to woodworkers, and a sensible compromise will be to use decimals to one place of millimetres. This will then give a saw set distance of 0·3 mm, which is approximately equal to 0·012 in.

from the lower wheel; (*b*) it may be raised or lowered by handwheel and lead screw to adjust for different length saws; (*c*) it has free movement up or down, with its downward movement restricted only by the support of a sturdy compression spring, and its upward movement restricted in turn by the constraint of the encompassing saw blade; (*d*) it has a tilting movement backwards or forwards controlled by hand screw or lever.

The first and second of these movements are self explanatory. The third movement is to ensure adequate stress of the saw blade under all conditions of work. As the saw gets warm due to cutting friction it expands, the spring takes up the resulting slack and reimposes the correct degree of stress. When cooling after use the saw contracts and the spring allows for this and prevents fractures. During work, hard or gummy spots in the wood may impose sudden extra loading on the saw; the spring compensates.

The mechanical devices by which springs are brought to play on bandsaw blades vary with each design of machine. Some use direct thrust under the wheel bearing, others have a form of lever and weight, with spring damping. All serve the same purpose, i.e. maintaining adequate stress.

A word associated with stress is *strain*. Stress can be defined for bandsaw purposes as the stretching pull on the blade. Strain can be defined as the stretching that occurs in the blade due to the pull, or stress. Stress and strain are directly related and are inseparable; where a blade is stressed it is proportionately strained. Popular usage in the woodworking industry gives the word *strain* most prominence.

Most modern machines include a strain gauge with calibrated indexing for each saw width. Narrow saws require less strain than wide ones and the gauge is set to indicate the correct strain for individual width saws. Some textbooks give strain tables in pounds, but these vary enormously, one book showing 60 lb for a saw 0·5 in wide and 22 gauge, whilst another gives 100 lb. In practice the built-in strain indicator is used only as a useful guide. Correct working of the saw is the only check. Should the saw vibrate and tend to rattle at the guides, more strain is indicated. There can be no indication of excessive strain other than a collection of broken saws over a period. Start low, and increase as appears necessary, is the rule. A working check is to spring the saw sideways with finger and thumb. Total, relatively easy movement should be no more than 18 mm. The fourth movement of the top wheel, that of tilt, is there more for the care of the saw and operator than as a practical aid to cutting.

Bandsaw wheels have flat, hard, rubber tyres. Saws can be positioned at the extreme front edge, in the centre, or at the extreme back. They will work well in each position. However, at the two edges, sudden movement (change in cutting pressure or direction by the sawyer) can force the saw off the wheels with resulting damage to the saw, machine and possibly the sawyer.

For these reasons then, it is the practice to run the saw at the centre of the wheels, and the tilt movement enables the control of this. Flexible belts used for machine drives tend to climb to maximum tension, and this phenomenon is utilised by running such belts on crowned (rounded) pulleys. Narrow bandsaws are semi-rigid and the wheel tyres are flat. The saw runs towards the lowest point, i.e. slides downhill, and wheel tilt enables the running position to be controlled at the approximate tyre centre. Tracking is affected by the degree of tension applied, and the saw track should be checked at every alteration of strain control. Tracking is done with the saw guides open and the wheels spun by hand.

SAW GUIDES

Although, in theory, saws should not be expected to cut what they cannot easily cut, in practice they often are. A sharp saw, in good condition, will cut its legitimate work with no rearward thrust at all. As the saw dulls, rearward thrust builds up. When knots and bad grain occur, the saw is again thrust to the rear.

To enable continuous work of reasonable quality to be produced, the rear of a narrow bandsaw is supported by a 'thrust wheel', and often two, one above the table and one below. These are free running wheels varying in diameter according to machine design, but are, on average, about 56 mm. They are positioned by the operator after the saw has been tracked and strained, and should be approximately 1 mm behind the saw. If the back of the saw is run hard against these wheels the back edge becomes work-hardened and crystallised, and fractures are likely to occur. The thrust wheels are part of the saw guide complement of the machine, the other parts of which are the top and bottom saw guides and the table mouthpiece.

Saw guides vary according to the machine design. They range from simple wood, fibre or metal blocks, to elaborate devices incorporating magnetic fields which are claimed to stabilise the saw. Very popular at the present time are those which use top and bottom pairs of steel rollers, pairs positioned face to face with the saw between, and static block types using hard graphite sections which are set to touch the saw lightly.

All guides serve the same purpose, i.e. that of assisting in the control of any sideways deflection of the saw blade. Knots may encourage the saw to run to the side; unskilled cutting of curves will do likewise. The guides, above and below the table, help to steady the saw and to keep it to a straight and narrow path. The final part of the guide system is the table mouthpiece, a round or square piece of hardwood, cut to allow passage of the saw. This block should be replaced frequently in order to keep as narrow a mouth as possible.

THE WHEEL BRUSH

To maintain a clean flat tyre on the lower wheel, and to avoid sudden repetitive jolts to the saw as dust builds up, a hard bristle brush is set close to, and permanently in contact with, the lower tyre.

THE BRAKE

This may operate on one or both wheels, and most often is in the form of a contracting shoe or band. Application of saw wheel brakes can be by hand lever

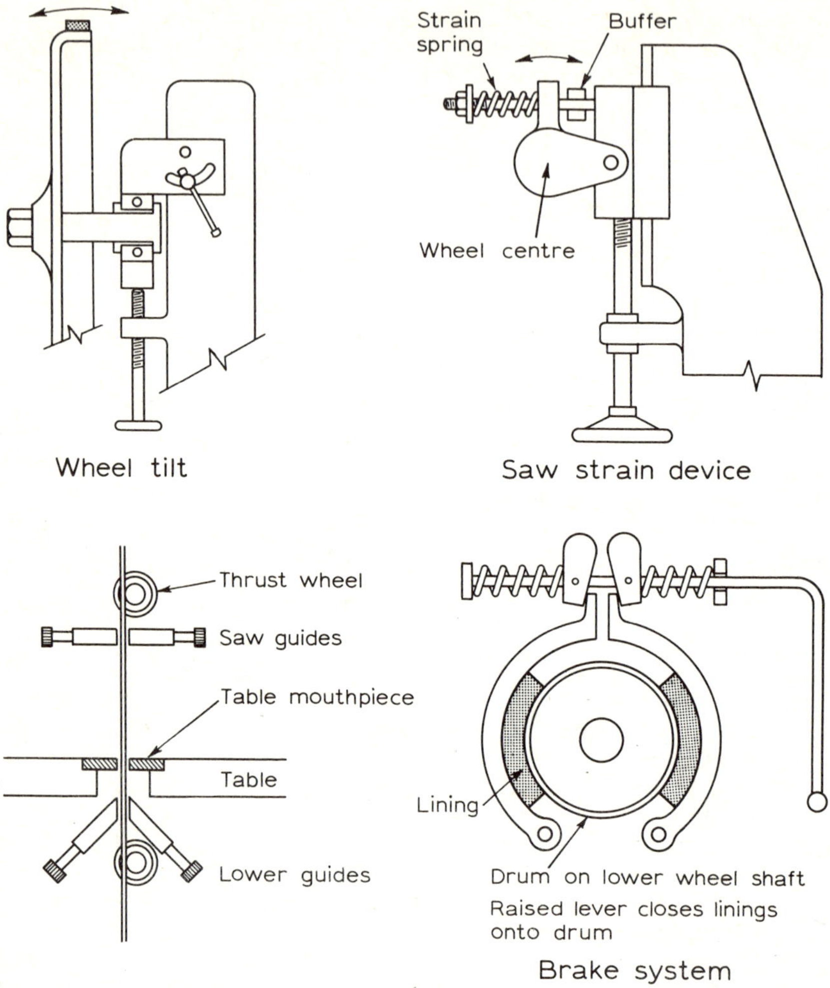

Figure 5.4 *Details of wheel tilt, saw strain, saw guides and machine brake. Design details vary, but the aim is to track correctly, strain, guide and stop the machine*

and rods, or foot pedal and cable. Sometimes operation of the brake lever will open the motor circuit and reset the starter for the next run. Figure 5.4 shows details of wheel tilting, saw straining, wheel braking and saw guiding devices.

Bandsaw speeds

Narrow bandsaws cut by the progressive removal of wood particles by successive teeth of the saw. Each tooth removes only that amount of wood that it can contain in its gullet for the period until it (the tooth) breaks free of the lower wood surface. Thus if the wood is fed too fast the saw chokes and deviates. If the wood is fed too slowly the teeth rub and lose their cutting edges, and this is often the cause and condition of those bandsaws used mainly by non-machine woodworkers. Too slow feeding of any machine tool produces rapid dulling of the cutting tool edge.

It can therefore be reasoned that any selected saw speed can only be a compromise, and in fact, of all machines, bandsaws vary most in their cutting speeds. The speed of 2000 m/min (lineal) is looked upon as being reasonable in practice, theory and engineering. This will vary down to 1500 m/min and sometimes up to 2200 m/min.

The calculation necessary to check for correct speed is the same as that for calculating the rim or peripheral speed of a circular saw. The motor speed, which, because of direct drive, is the wheel speed, is read from the motor plate.

Example 5.1

Find the lineal speed of a narrow bandsaw (in m/min) running on machine wheels of 762 mm diameter, at an indicated motor speed of 825 rev/min.

$$L = \frac{\pi DR}{1000}$$

where L = Lineal speed of saw (m/min)
D = Diameter of wheels (mm) and the divisor of 1000 converts millimetres to metres
R = Indicated motor speed (rev/min)

Hence $$L = \frac{3{\cdot}14 \times 762 \times 825}{1000} = 1973 \text{ m/min}$$

Another calculation necessary is that relating to the correct length of a bandsaw. Shop-floor practice is to wind the top wheel up to its highest, lower it about 20 mm and then to run string round. This would give a suitable working length and allow for final strain. The length would then be transferred to the shop floor, there to be marked permanently.

However, examination candidates must answer questions requiring the calculation of saw lengths. The length of a bandsaw will be the circumference of one wheel plus twice the wheel centre distance. Prior to measurement of this, the top wheel will be marginally lowered. The formula is:

$$L = \left(\frac{\pi D}{1000}\right) + 2C$$

Where L = Length of saw (m)
D = Wheel diameter (mm) and the divisor of 1000 converts millimetres to metres
C = Wheel centre distance (m)

Example 5.2

Find the appropriate length in metres for a bandsaw to fit a machine having wheels of 762 mm diameter and a centre distance (allowing for strain) of 1·5 m.

$$L = \left(\frac{\pi D}{1000}\right) + 2C$$

$$L = \left(\frac{3{\cdot}14 \times 762}{1000}\right) + 2 \times 1{\cdot}5$$

$$\therefore L = 5{\cdot}39 \text{ (or } 5{\cdot}4 \text{ m to 1 place of decimals)}$$

Power

The power used is proportional to the depth of cut, but many other factors apply. A standard 915 mm machine will need a motor giving an output of 2 238 W.

6 The Spindle Moulder and Shaper

The name

The spindle moulder takes its name from the central spindle that does the work and from the fact that its purpose is moulding and shaping.

Like the router, it is a machine that is basically simple but that has an enormous potential that must be realised by clever application and adaption. In small shops, the spindle is usually used as a general purpose moulder with perhaps adaption for the stair stringing, dove-tailing and some tenoning. In larger shops with a wider range of machines, the spindle probably remains as a spindle with comparatively straightforward moulding and shaping as its main job. It is in the larger furniture shops that the full potential is exploited, for on large scale repetition even small insignificant cuts cost money if done by hand—thus complex jigging is often devised for very simple cuts. The jig is necessary to hold and present the work at any peculiar angle required.

The simple spindle has its 'big brother'—machines that mould and shape on automatic cycle—that once set with the correct jigs and cutting tools, operate for long periods producing accurate work and require only that they be fed. These larger machines that stem from the spindle are called *shapers*, for most often they are used to produce shaped components rather than the mouldings and moulded edges of the spindle.

The purpose of the spindle is to mould, shape, cut and form in any way that can be devised. Included in this capacity are straight mouldings and rebating, stopped grooves and moulding, returned ends, tenons, scribes, housings (dovetailed and square), stair stringing, notching, corner jointing, square turning, reeding and fluting, panel raising and edge jointing—a full list but by no means comprehensive. Figure 6.1 shows a varied range of spindle work.

The machines

The machines in the spindle range are the standard spindle moulder which is a box-like flat-topped machine with its single spindle projecting upwards through its flat table. On this table are attached the standard work fences that guide the work into and out of the cut. Other attachments, such as feed units and ring fences, are also bolted to this table. Below the table are the machine bearings

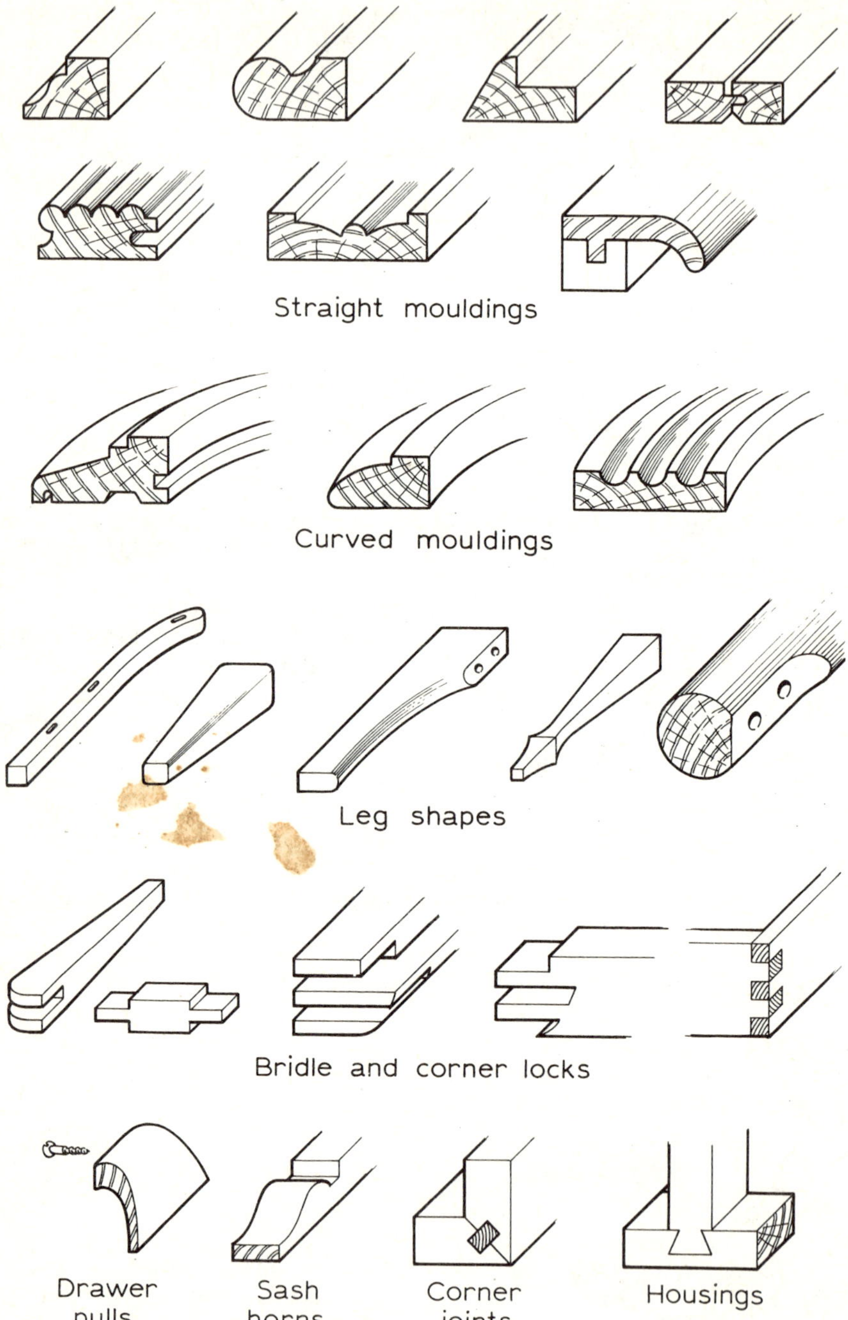

Figure 6.1 *Range of spindle work. The straight and curved mouldings are fairly straightforward but the leg shapes require jigging or some form of static fixture, as would most of the joints shown*

and spindle slides, together with the spindle rise and fall mechanics, drive motor and brake and lock units.

The machine is made in various sizes and with variable preselected spindle head speeds. There are comparatively slow runners and there are high-speed models. Some are described as 'heavy duty', and these have slower speeds and heavier castings, and can cope with the largest cutting tools.

Next in the range are double spindles, having all the appurtenances of the standard model but having them double so that through the one, flat, work table two spindles project.

Developing still more there is the automatic chain-feed spindle, where the single cutting spindle is set through a ball-bearing collar and sprocket ring. The collar acts as a roller guide to jigs which are fed to the cut by the sprocket ring

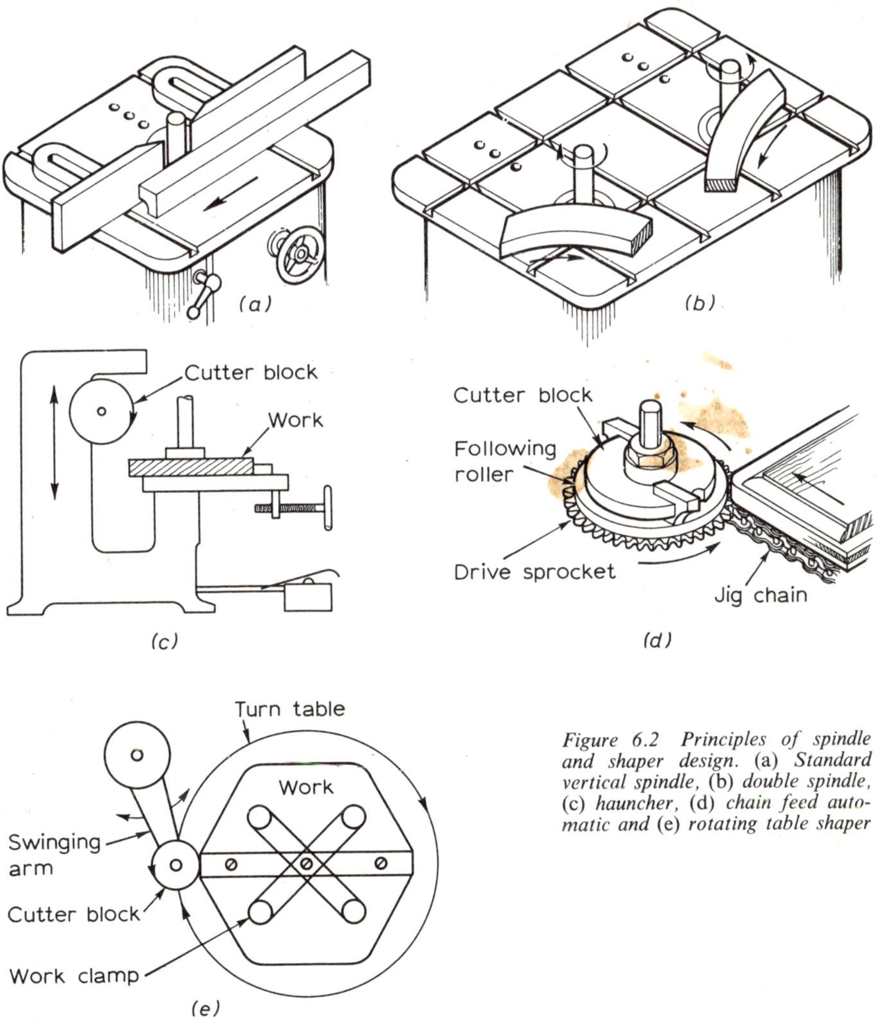

Figure 6.2 Principles of spindle and shaper design. (a) *Standard vertical spindle,* (b) *double spindle,* (c) *hauncher,* (d) *chain feed automatic and* (e) *rotating table shaper*

which engages in a roller chain set in a groove around the base of the jig. When the jig is loaded with work it is fed so that its chain meshes with the drive sprocket ring. The ring, rotating, feeds forward the jig whose path is controlled by the roller collar that runs in contact with the jig templet edge. So it is that repetition shaping is performed without hand guidance or skill.

The big brother machines are not called spindles. They go by the name *shapers*, with the exception of one that is known variously as the *horizontal spindle*, the *notcher* or the *hauncher*. The horizontal spindle is not a moulder or shaper, rather it is a general purpose cutting tool that leaves its use to the operator. Basically it consists of a large rectangular casting that stands on end. Projecting from the work side is a horizontal table that carries any fences or jigs necessary for work positioning. Behind the work table, and across the machine front is a horizontal spindle. On this are mounted the cutting tools which are most often standard spindle-moulder tools. Work is positioned, clamped and the horizontal spindle is brought down, passing the work table and cutting the workpieces which are projecting into its path. In this way corner notching for shelves, corner locking for boxes, complex mitre joints, rounded mirror parts, small tenons, hinge recesses, leg tapers and square turnings can be cut.

Increasing in scope and size again are the *shapers* whose work range includes chair arms, rails and legs, table top profiling, chair seat sizing, tennis racquet press jointing and shaping, window sash sizing and edge rebating, brush handles, gun stocks, and table leg and rail shapes.

The machine principle is based on a vertical spindle, the lower end of which carries the cutting tool. The spindle is carried on a pivoting arm which is free to swing towards, or away from, the workpiece which is carried on the large rotating table. The workpiece is positioned on top of a shaped jig which has a forming templet as its base.

Beneath the cutting tool there is a ball-bearing roller collar, the outer periphery of which moves into contact with the jig templet and follows it during the cut. Thus it is that as the table rotates the various contours of the jig templet cause the swinging arm to move in or out and shape the work. The workpiece, positioned above the templet, is cut by the cutting tool that runs above the ball collar. On the larger machines there are two swinging arms and cutter heads. Figure 6.2 contains outline diagrams of each of the above types of machines.

The standard vertical spindle moulder

The main feature is the spindle itself. This is carried in substantial bearings and is most often driven by an endless short-centre flat belt, from a motor built into the machine main frame. Rise and fall is by bevel gear, or worm and pinion. The top end of the spindle is arranged to carry varying cutter-head spindles, the most common of which are the French head spindle and the plain spindle, the latter having a threaded portion at its top end. Some operators call this the *English* head. Other cutter-carrying fittings are chucks to take dove-tailed and recessing cutters, and stub arbors that carry flush top cutter heads.

The method of fitting for these heads varies with machine design. Some have keyed shafts, some tapered, and some have collars that are held down by a locking ring. All methods are effective, and interchange takes only minutes.

The spindle must always have a lock to facilitate interchange of spindle heads or cutting heads and this is usually hand-lever controlled and most often takes the form of a yoke that engages flats on the spindle shaft.

Brakes are fitted to enable quick interchange and adjustment and as an emergency method of stopping the machine if accidents should occur. Brakes are mostly friction pads that are pulled or pushed against a smooth ring on the spindle shaft. Some machines have brakes that interlock with the electrical circuits and so act as stopping control.

Fences are guides for the work and can be fitted to the table top as two pieces, before and after the spindle, or they can be incorporated into a complex fitting that carries all adjustment and lock screws as well as guards and dust hoods. The standard method of fence unit fitting is to have dove-tailed slots across the table into which fit dove-tail-headed bolts. An alternative method is to have threaded holes in the table top to receive bolts.

Every fence unit has screwed adjusters to allow alignment or out-of-alignment if necessary. They also have a built-in facility to slide towards the spindle or out and away from it. Where the two fences are in one unit this is quickly set but can prove bulky for some jobs. Both methods are effective. The separate-fence system is keyed to the table slots for alignment and it is not possible to slew such fences round to an angle with the table edge, as it is with the fences that are only bolt held. Setting up is probably more quickly done if fences are in two units and for some jobs one fence only can be used. Figure 6.3 shows various standard spindle details.

DRIVE DETAILS

Drive is mainly through short-centre flat belts from built-in motors. This method gives a choice of speeds, for the motors can have a two-step cone pulley, each step being of a diameter necessary to give the correct ratio to the spindle pulley to produce the correct running speed. A two-speed motor thus gives four-speed selection.

Motors are usually of the fully-enclosed squirrel-cage induction type, and these give one revolution per cycle of current alternation for one pair of poles. With four poles (two pairs) only half of one revolution is given per cycle. At the standard 50 Hz (cycles per second), a motor will run at approximately 3000 rev/min with two poles, and at 1500 rev/min with four poles. By switching, either speed is available. If a reversing switch (which changes over the connections of two wires and thus reverses the directions of the stator rotating field) is available, four forward and four reverse speeds are obtainable. Motor principles are detailed in Chapter 7.

The actual speeds are built-in to order, for it is simply a matter of pulley diameters. In general, larger cutter blocks must run more slowly than smaller ones. The aim is to maintain as high a cutting speed as possible whilst not taking liberties with balance. Blocks in near perfect balance such as solid-profile heads can be run, diameter for diameter, faster than square cutter blocks on which loose cutters are bolted, and where dynamic balance is always suspect.

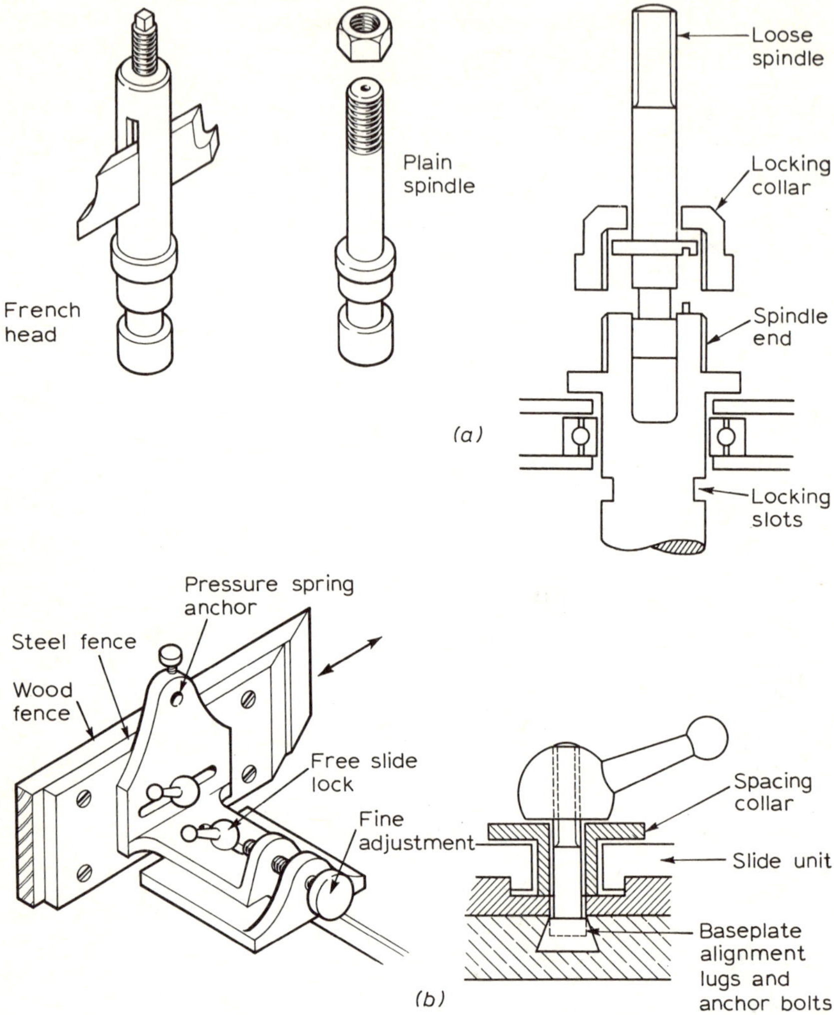

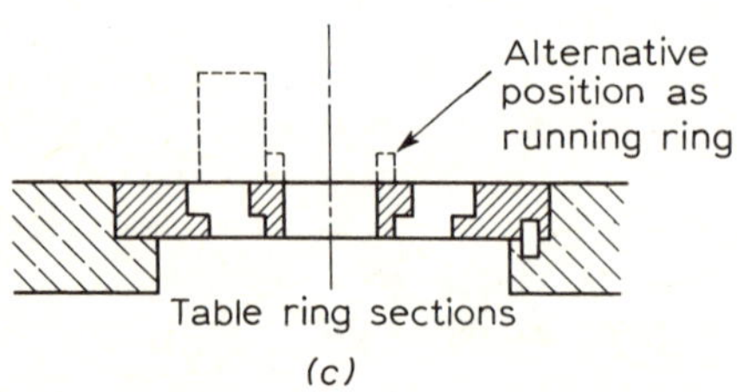

Figure 6.3 Types of spindle heads. (a) *French spindle and plain spindle; the former is a spindle and cutting head combined that only needs cutters to be fitted, whereas the latter simply receives any form of cutter or cutter block that has a centre hole. The Wadkin method of head interchange is shown in the top right corner.* (b) *and* (c) *are a fence unit and table rings*

With shaped work, where consequent change in grain direction is present, high speed is essential. This brings with it limitations in cutting diameter, which brings in the necessity for multi-edge profile blocks to keep up the number of cutting edges getting into the cut. Every practical machinist knows that a square cutter block and cutters cuts more easily than most other cutting tools, but at the same time recognise the dangerous nature of such a tool and the limitations in achieving fine balance. Thus the smaller, but faster, profile head is accepted. The small high-speed spindles run as fast as 15 000 rev/min, but are restricted to small diameter, special cutter heads.

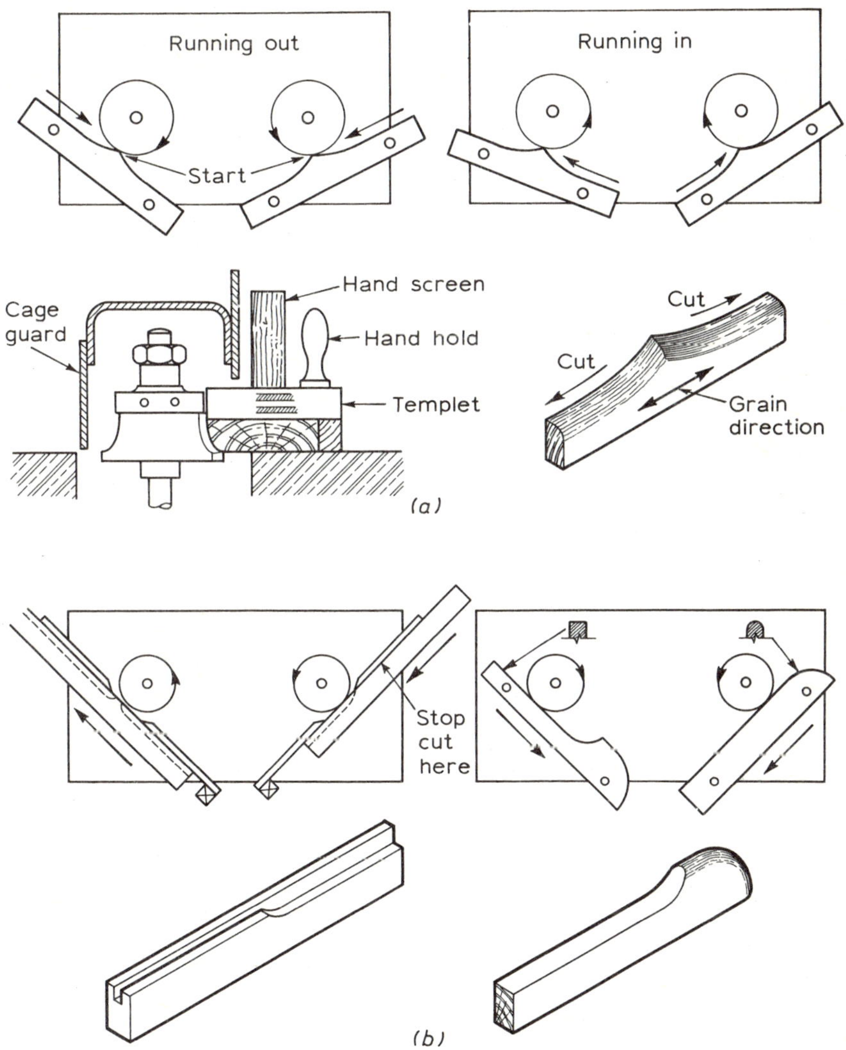

Figure 6.4 *Double spindle (mainly used for shaped working).* (a) *Run-in and run-out working methods and* (b) *alternative types of work*

Double spindles

These have all the fittings of the standard machine but have a large work table common to both spindles. The idea behind this design is to enable shaped jobs to be cut with cutting always down the grain. This means that for a straightforward cleaning-up job after a shape has been bandsawn, both spindles will be set with a circular safety block and eccentric ring fence, or both will have ball collars instead of the ring fences. The spindles will be set to run in opposite directions. They can run-in or run-out, which means that the cutters will run-in together, from the front of the machine, or they can run-out, again looking from the machine front. The shape of the job can influence the choice, but it is nearly always most safe to run-out, as, in the event of a kick back, work and hands are knocked away from, and not towards, the other spindle.

Double machines can be used for moulded rails with curved edges. Both heads are set with identical profile cutters, but each to the opposite hand. Work is commenced on one head and finished on the other, the change being effected where change in grain direction occurs. An alternative use of double spindles is to set both with different cutters for use where one piece of work would require two sections, such as above and below middle rail in a half-glass door. Both heads could run in the same direction and handling time would be saved. For extra-heavy cutting one head could be set to 'rough' and the other to finish. The machine can also be used as though it were two machines. Each head would be set with its own fences and have its own operator, and a board screen would be erected between. Figure 6.4 shows the principles of design and use of the double spindle.

The automatic chain feed spindle

Where production is repetitive, batch produced, or in very long runs, automatic feed is desirable. Operator error is eliminated as is much of the danger associated with fast cutting machines. Accurate repetition is built into the jig and cutter set-up. With automatic working, workpieces are loaded to the jig by hand, and thereafter cut without hand control.

THE METHOD

The spindle base is heavier, and the working table much larger, than for the conventional machine. There is a heavy-duty spindle with usual rise and fall, lock and brake, and a more powerful drive unit. There are no fences. The cutting head runs much higher above the table than is normal.

Around the base of the spindle unit, where it emerges from the work table, there is a sleeve. Running on this is a sprocket with teeth at approximately 15 mm pitch. This sprocket ring is driven from its own motor and runs at a selected controlled speed. Its rotation is clockwise viewed from above. Above this sprocket and mounted on the same sleeve there is a free-running ball-bearing collar. Above this again is the cutting block mounted directly to the emerging spindle. The cutting tool used can be of any standard type suitable for the job. The spindle runs the same way as the sprocket, which is the reverse to that normal to spindle moulding. In front of the cutting block, or sometimes surrounding it, there is a work hold-down pressure pad, either spring or hydraulically loaded.

The jig can be shaped to cut either 'one off', or a large plan area is contoured to cut several pieces of work at once. If one piece only is to be cut, the jig contour approximates to the job outline in shape and size; chair-seat all-round shaping is one example, and small jobs are usually distributed around the periphery of the jig, and relative portions of the jig have the correct plan shape for this.

Workpieces are positioned by small stops and location pieces and held by small toggle clamps and the top down-pressure pad. Large one-off shapes, chair seats and table tops, etc. that are to have full contour shaping are usually held on sharp pins as with router work.

The jig is made in three sections. The base section, of thick multi-ply, has its centre cut away to form a large circular hole. The outer edge of this section is rebated to take a 15 mm pitch roller chain, the chain being built with metal tags for the purpose. The middle section, again of multi-ply, is fitted on top of the base, and has its outer periphery smooth and contoured to the plan shape required. This section is called the *templet*. Above this the working top is fitted and this has the various location stops and cramps fitted to it.

In front of the spindle head, and radiating from it, there is a slot cut in the work table. Through this projects a short arm carrying a ball-bearing roller above the table surface. The lower end of the short arm is connected to a double-acting pneumatic cylinder operated from a foot-pedal control. The jig centre hole drops over the ball roller. To bring the jig to the 'load' position the foot-pedal is used to cause the air cylinder to pull the jig away from the spindle. Small jobs, set around a large jig, are changed over as the jig is working and as they pass the operator.

When work is to commence the cutter block is started and the air cylinder is used to push the jig into position so that the roller chain meshes with the rotating sprocket ring. Work is held down by the top pressure, the jig is fed onwards by the sprocket and chain, its path controlled by the ball roller and templet edge, and the job is cut to section and contour.

To ensure clean cutting and to avoid break-out on corners 'climb-cutting' is used. This means back-cutting, where cutters and work move in the same direction at the point of contact. Only with machine-controlled feed and work holding is this method possible. Its advantage is that waste material is cut away before cutters reach bottom-dead-centre of the cut so there is nothing left to break out as the cutter leaves the cut. Cutters used in this way wear slightly more quickly than is usual with conventional cutting as the cutting edge is held more closely in contact with the wood and more friction occurs.

A peculiar result of this chain and sprocket method of feed is that at the point of most risk to the work (sharp corners of tight radii) the feed speed slows down. This is not necessarily intentional but is inherent in the geometry of loci. The extent to which slowing occurs depends on the relative diameter of the cutting circle, the job curve or corner, and the jig curvature. The chain way and the templet are usually closer in to the centre of the spindle head than is the flying circle of the cutter. Imagine a perfectly circular job; its circumference will always be less than that of the jig outer edge, and as a smaller circle, will always travel less in distance than the jig edge. Thus the jig feed speed will not be the same as that of the work. Extend the principle to the situation where the jig corner radius is possibly 25 mm and the cutter overhang is about 25 mm, then the cut corner will be square in plan and the cutter will hold there until the jig has fed about 20 mm.

The chain feed spindle is suited to brush work, chair components, table tops, sash sizing and any job of sufficient quantity whose requirement is for edge profiling and shaping subject to there being no acute internal corners.

The same principle can be used for edge sanding and the spindle head is

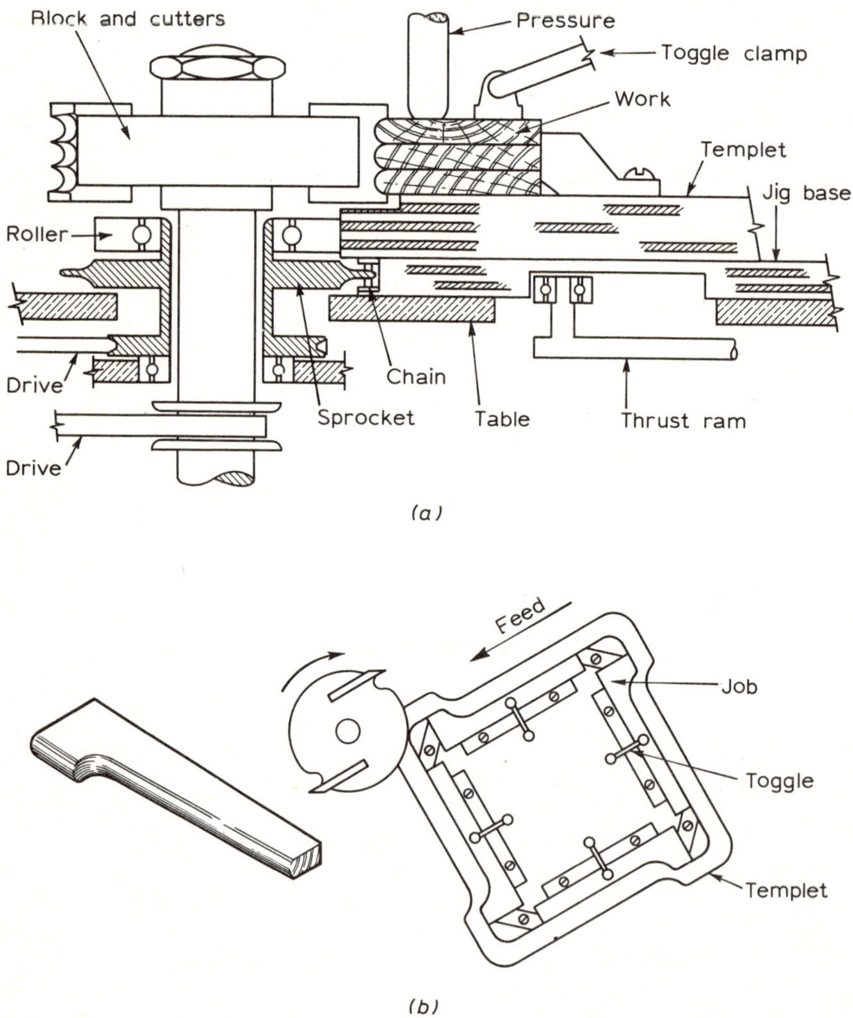

Figure 6.5 Automatic chain feed spindle. (a) *Sectional view and* (b) *multi-station jig*

replaced with a belt guide and backing pad, whilst a narrow belt runs across to sand work. Full details are shown in Figure 6.5.

The horizontal spindle or hauncher

The cutter spindle of this machine is horizontal and is mounted in bearings that are able to slide up and down machined slides for a distance of 240 mm. The bearings are close together, with the spindle shaft overhanging at both ends. At one end is the driven pulley, and at the other, the machine cutting head which can be any standard spindle cutter head. The spindle projects far enough to take blocks up to 160 mm long.

In operation the spindle rests at the top of its stroke and moves down in a straight line when the foot pedal is depressed. The workpieces are positioned against a fence and held flat down by an air clamp that operates prior to the downward movement of the spindle.

The feed stroke is powered and controlled by a hydro-pneumatic system. This consists of a double-acting pneumatic cylinder that has its speed controlled by the flow of oil through restrictor valves. This gives easy control of travel, and what is more important, smoothness of travel. Air alone can produce jumpy feed, as resistance in cutting occurs.

The cutting stroke cycle which is 'down/up' is controlled between 3 to 4·8 s and stroke length is varied from 120 to 240 mm.

The work table, positioned across the front of the machine, has adjustment towards, or away from, the cutting spindle. It carries the location fence which

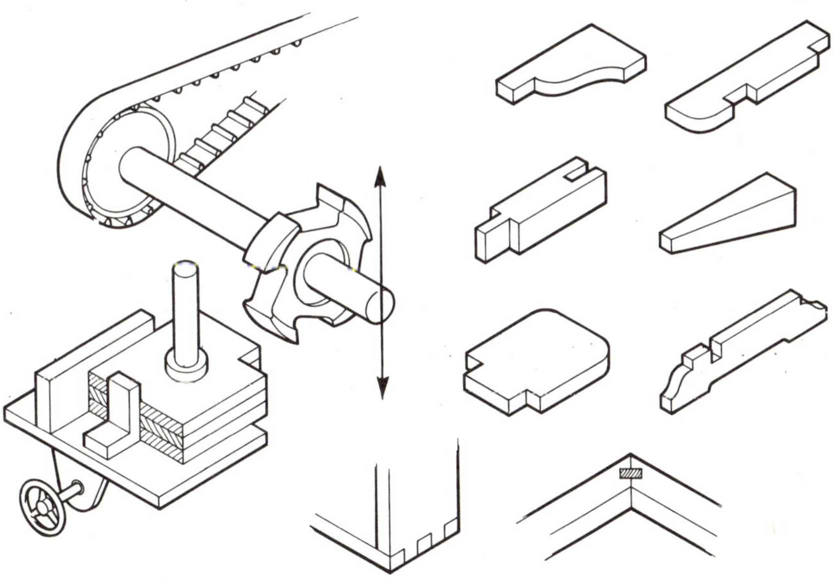

Figure 6.6 Horizontal spindle or hauncher. The work range depends on the cutters available and the ability of the methods team

may be pivoted, and any jigs necessary for correct presentation. Work range includes corner locking, corner notching, corner rounding, rail tenoning and haunching, half lapping, housing, scribing and dove-tailed tenoning. With suitable jigging, small leg tapers can be cut.

A double-ended version is available. A unit is positioned each end of a track slide with spindles parallel to each other. One, or both units can be pivoted in plan through 90° to suit complex jobs. Adjustment along the track controls shoulder length of double cut pieces.

The 'Hauncher' is made by J. K. O. Cutters Ltd. Examples of work range and diagrams of design principles are shown in Figure 6.6.

The rotating table shaper

Often called the *turn-table* this is the largest of the spindle or shaper types of machines. It takes its name from the large diameter rotating table that carries the work. The table size varies from 710 mm to 2·150 m diameter.

On the rotating table there is a multi-ply jig. Various portions of the outer contour of this are shaped, in plan, to identical shapes required on the finished workpieces. The shaped portions are blended together to form a complete contour which can be followed by a ball-bearing following roller.

This outer contour of the jig can be made up of several individual shapes, matched to suit short workpieces positioned around its periphery, or it can be identical with the total plan contour of one large job that is to be machined fully around its edges, i.e. shaped. Two table leaves can be positioned back-edge to back-edge to have their remaining three edges shaped. The two leaves would then form one large fully-machined shape. When several short pieces are distributed the jig is called *multi-station*, and where full machining of one piece is to be performed, the jig is *single-station*.

Various combinations of work stations are possible. Chair back legs, which are approximately arc shaped, have to be cleaned and shaped on two edges, i.e. inner and outer. These are arranged on the jig four at a time. The first piece will have its convex edge to the cutter, with its sawn back edge against location blocks. Following this will be a second piece, which is arranged with its concave face to the cutter whilst its other convex edge, which has already been planed, is positioned against precise locators. When this piece is past the cutter it is finished and is taken from the machine. Following this second position there is another convex shape which has in it a leg shaped on its convex edge. This is removed and replaced with a band-sawn leg. The half-cleaned leg will now go into the concave position which is number four on the jig. Thus each revolution of the jig will complete two legs.

With smaller pieces that may require shaping on all sides (using 'sides' loosely, as is necessary when generalising about shapes), four pairs of work stations can be built in, and the operator loads station *one* of each pair with a work blank and repositions each pair into station *two* of each pair when it returns to him. In this way four pieces per revolution would be produced.

Compressed-air clamps hold down each workpiece, and these are operated automatically releasing and clamping at preset positions as the table rotates.

Adjoining the air clamps are spring-loaded steadies, whose job it is to hold each workpiece until it is far enough away from the operator to avoid risk to his fingers from the air clamps. For single-station work such as sash frames and table tops, one large centre clamp is used, and this is beam mounted to leave the whole area of the table clear. Rotation of the table can be retarded at critical points, heavy cuts or sharp directional change positions, and this is controlled by a mechanical/pneumatic device which cuts table speed to half over limited distances.

A large cam disc is set with striking cams which open and close an air valve as they pass. The valve allows compressed air to flow to a pneumatic cylinder that opens the halves of a variable-speed 'V'-belt pulley. The belt forms a different ratio to its driven pulley on the table rotator and the table slows down. As soon as the critical point is past, the table returns to the normal cutting speed. An alternative design for the retardment of the table has a double friction clutch. Each clutch controls a pulley of different size. One clutch is always engaged. The striker cam simply throws out the normal speed pulley and engages the slow speed pulley for the duration of the 'slow' period and then allows the 'fast' pulley to re-engage.

A friction brake closes automatically and power to the rotating table is cut off if the air pressure falls below the safe working pressure, which is usually about six times the atmospheric pressure. In the Imperial system 1 atmosphere (1 atm) is equivalent to 14·7 lbf/in^2. The S.I. equivalents of 1 atm are $1{\cdot}013 \times 10^5 N/m^2$ (or 101·3 kN/m^2) and 1·013 bar. Hence 1 atm $\simeq$ 1 bar.

Table speeds can be selected from 15 s/rev up to 120 s. Table rotation can be set to continuous, periodic or full manual control.

Air drills can be fitted to the table and these drill components whilst they are being shaped. Sanding heads can be fitted to operate behind the cutting heads.

THE CUTTING HEADS ON SHAPERS

Small machines have only one cutting head but large machines can have two. With two heads, chair back legs can have their two edges cleaned and shaped (one head) and have a short top portion bevelled to receive a moulded chair back. Two templets are necessary when using two heads, and one is placed on top of the other. Each head would have its own following roller set at the correct height for its own templet. One head would be set with a square cleaning profile, whilst the other would be set with a double bevel profile. The double bevel is necessary, for chair back legs must be produced in handed pairs, and to machine one pair at a time necessitates that they are machined together, i.e. face to face. Each station would be loaded with two sawn blanks (accurately thicknessed) and both would be cleaned, shaped and bevelled together.

The following roller for the square cutting head would allow the head to be in continuous contact for the full length of the leg, whereas the roller of the second bevelling head would move in and out allowing contact only for the length of bevel required.

The cutting head units are carried on swinging arms that are held with their following rollers continuously in contact with the templets by compressed air.

If air pressure falls the arms move away out of cut. At the lower end of the head unit there is the cutting block, set with any form of block, but which are most often of the slotted collar type, or circular safety design. Below the cutting block there is a bracket on which runs the ball-bearing following roller. The roller bracket can be adjusted by a fine screw to allow increased or decreased cut.

To ensure clean cutting under all conditions, the cutters climb cut (back-cut). As the rotating table travels clockwise in plan this means that to climb cut the heads run anticlockwise. Coupled with climb cutting, the heads run at very high speeds, and for this they require frequency changers as the cutting heads run directly coupled to their drive-motor shafts.

Climb cutting, high speed, and table slowing at points of risk, mean that most jobs can be cut to an almost non-sand surface. However, one common error, due mainly to operation and not to machine design, is that of chatter, which produces ripples at heavy cutting points. This indicates that the cut is too fast, too heavy, or that the jig contour is changing too rapidly.

Safety in design of shaping machines

As the operator stands directly opposite across the circle from the cutting head(s) he is out of the proximity of danger. The machine's built-in safety features protect both itself and persons in the near vicinity.

Should the air pressure fall below the required level to maintain contact between the following roller and templet, and below that required to operate air clamps, the swinging arm moves away and the power is cut off; the table brake is also applied. Cage guards surround the cutters, and the air clamps are timed to operate after each work station has moved on from the loading point.

Figure 6.7 shows details of single- and multi-station work, using single and double heads. The head, follower and templet are also shown, together with a diagram of one table-retarding system.

Spindle and shaper cutter heads

FRENCH HEAD

Most common of all spindle cutter heads is the French head, although its use is declining. It consists of a plain vertical spindle having a vertical slot cut through it for part of its length. A set screw is threaded down from the top of the spindle and this grips any cutter that is fitted through the slot.

The cutters are cheap and easily made. The material is usually bought in black strip, but can be bought in ground stock lengths. The strips may be straight high-carbon steel, or tungsten-alloy steel, which is harder and longer lasting in cut. Both types can be hardened and tempered to give cleaner cutting with longer life.

Cutters cut through the centre line of the head and have no cutting angle; they meet the wood at 90° at top dead centre (maximum depth of cut). This allows cutters to be made exactly to the opposite of the mould section required. If the section has a circular arc profile, the cutter is the same.

Cutters are made by grinding. Correct length, which allows for depth of cut, balance, clearance, diameter of spindle and for regrinding if necessary, is cut off by first nicking both faces with a hacksaw and then snapping in a vice. Cutters are ground by presenting them free-hand to the grinding wheel. As the profile forms it is compared with the section required that is drawn on paper.

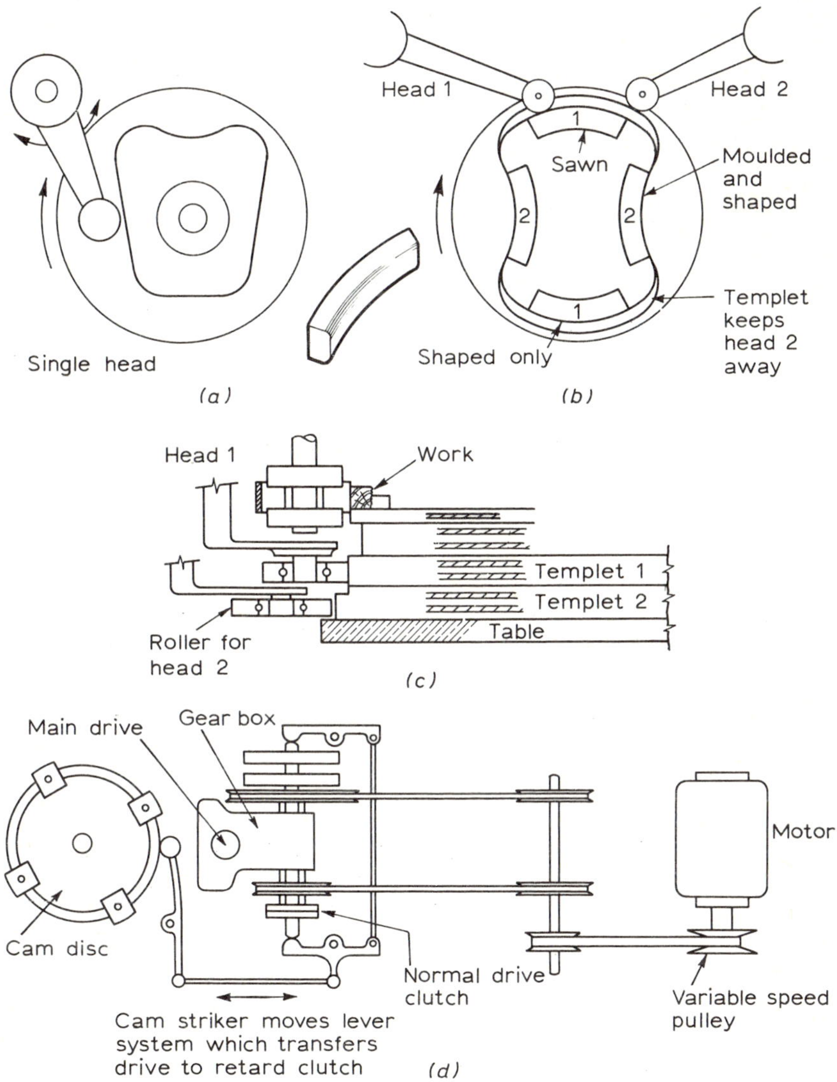

Figure 6.7 Rotating table shaper. (a) *Solid chair seats or frames may be shaped on the rotating table machine using one head and a single station work system; with two heads* (b) *multi-station jigs both mould and shape;* (c) *relationship between head, follower, jig and work;* (d) *Rye system of table control*

Small beads and sharp corners are filed. Clearance behind the cutting edge is kept to 40–45° and to 5° behind the side or horizontal edges. The exact grinding angle will vary according to the nature of the wood to be cut, the cutting life required, and the quality of the steel in use. Softwoods may be cut with sharp thin edges, but hardwoods require thicker, tougher edges.

For reasons of safety, cutters are made to cut below the work if possible. This ensures that if the work jumps while being fed, a lump is left, which can be removed later. If the cutter is cutting over the top, a jump will spoil the job. If the cutter is underneath, the job protects the operator and if the cutter should break or fly out, the work is there again to protect the operator. Twisted wood is far less likely to be snatched and thrown back if the cut is underneath.

The working method depends on quantity and size of cut. Short runs can be picked out, each moulding member being machined separately with its own simple cutter. Heavy cuts can be made in several bites, each setting getting deeper.

If the job is longer running, repeated cuts would make it too expensive, so cutters are made to full profile, or even made to cut at both ends. Two or more cutters can be put into the spindle slot at one time and this simplifies the making of complex cutters.

At all times a non-cutting packing piece should be inserted on top of the cutter(s). This spreads the load of the screw and avoids damage to the top edge of the cutter.

Because cutters have no natural cutting angle, they tend to scrape and so they are sharpened in such a way as to create a cutting angle. They are ground to shape and then made sharp, using a swiss file or abrasive slipstone. Swiss files have very fine cutting teeth and are of a suitable slim section. After the edge has been made sharp, a burnisher is used. This is a very hard steel rod, roughly elliptical in section with one edge very thin, but rounded. Burnishers are made from swiss files; all teeth are removed by grinding and then the steel is highly polished.

The cutter, held face side up in a vice, has its extreme face edge highly polished by hard pressure from a burnisher. This is to work harden the edge and to iron out micro-scratches left from the file. The burnisher also turns the cutting edge slightly downwards. The cutter is then repositioned in the vice, approximately vertical with its cutting edge upwards, and the burnisher is drawn hard across the upturned edge, at an angle just clear of the grinding angle. The sharp cutting edge is thus turned over, and now has a hook on it. This hook is not a burr, which by definition is a jagged torn edge. The hooked edge will now cut, and not scrape, as it now inclines towards the oncoming wood and has a small cutting angle.

The French head is used for straight or curved work. Most often jigs are not necessary, for the job can be run directly in contact with the rotating head.

The square block

Square cutter blocks are used on many machines. Their use is dying out in large shops because modern profile heads are becoming less expensive. A collection of profiled heads can be used to pick out and shape most common sections and when this type of head is in use much cutter grinding and setting is saved.

Square cutter blocks cut more easily and freely than any other type of block; they have weight, speed and clearance, all of which make for quick, clean, cutting. They are also dangerous, for their irregular outline is difficult to guard, and the long projections of the cutters make snatching common if there is any uncontrolled movement of material when in the cut. Guards and safe working methods will be discussed later.

The square block ranges in size from about 60 mm square to about 110 mm square, and in length (height on the spindle) from 50 mm to 120 mm. On each face of the block is a dove-tailed groove into which slide the heads of the cutter bolts. Heavier blocks for four-sided moulders often have 'T'-grooves which are considered a better engineering proposition because there is no wedging action to cause distortion; for practical purposes this has little significance. These heavier blocks often have faces ground slightly hollow in order that cutters can be pulled down to a snug fit to the leading edge of the block thus assisting in preventing chips wedging under the cutter face. The hollow also serves to prevent cutters moving from their preset positions whilst holding nuts are being pulled tight. This degree of design engineering must be paid for, and so most square blocks for spindle moulders are of the flat-face dove-tailed slot type. For fast planing on four-sided moulders square blocks have 'lips' (see Chapter 2). For spindle work (mainly moulding) these lips have no value.

Into the slots on the block faces the cutter bolts are fitted, the size of which, when pulled tight down on the cutters, must be such that the inner bolt-head face is below the block face. Bolt-head faces above the block face prevent the cutters from seating firmly, and lead to flying cutters.

Cutters are either mild steel, or, for better quality, tool steel (medium to high carbon). These are faced either with a tungsten alloy steel or with High Speed Steel (h.s.s.) which is a special alloy of tungsten, nickel, chromium or cobalt and carbon steel. These metals are all hard, and h.s.s. is tougher and longer wearing than straight tungsten alloyed with carbon steel. Toughness, as an engineering definition, means the ability to absorb blows and work, and tough material tends to be softer than hard material which tends to be brittle. H.S.S. exhibits an apparent contradiction in that it is both hard and tough, although both these terms are relative (tungsten carbide is exceptionally hard and also brittle).

The facing material can be carried along the whole of the cutter face, or it can be inserted as a short section to cover the cutting portion of the cutter. Short facings are usually brazed on as separate tips, but long facings are usually fusion welded and rolled during the cutter billet manufacture.

Cutters are thick, ranging from 9 mm to 10·5 mm. For heavy-duty work they can be extra thick, up to 12·5 mm. So that they can be attached to the block, they are slotted, the slots being open-ended for spindle work, or, if called for, closed slotted for four-sided work. Bolts (and nuts and washers) should be identical; odd bolts cause unbalance.

To improve practical setting some operators use paper packing between cutters and blocks. Its use varies from thin full-face paper, to thicker strips across one, or both ends of cutters. Reasons for use are two-fold: paper increases the initial friction between cutters and block and prevents movement whilst tightening, and paper under the back end of the cutter closes the cutting end of the cutter 'snug' to the block face. Both reasons are valid when cutters are old and possibly distorted. New cutters and blocks with good faces need no packing. However, as packing is such common practice it is accepted as the thing to do. Packings should be thin paper, and if not used under the whole of the cutter face should be used only under the back end of the cutter. Packings of thick paper distort cutters.

CUTTER PROFILES

Square block cutters have a natural cutting angle which varies according to the size of the block and the projection of the cutter. Increased projection on the same block decreases the cutting angle. The same projection on a smaller block decreases the cutting angle, and any decrease in projection increases the cutting angle.

It follows that no two points of different projection on the profile of any shaped cutter will have the same cutting angle and the same relationship with the work. A cutter removed and re-used on another size block will develop new cutting angles. Thus it is that mould profiles cut by identical cutters used on different size blocks, will differ. This variation in performance is due entirely to the change in the cutting angle. The cutting angle is the angle contained between a line drawn from the tip of the cutter to the block centre and a line drawn along the cutter face. If looked at in plan this means that at top dead centre (cutter cutting at maximum depth) the face of the cutter is inclined towards the approaching wood and is not at 90° to it as it is with the French head cutter.

If, on one block, there are two pairs of cutters, one pair planing and one pair grooving to produce a groove 12 mm deep, the position would be that the planing cutters would be set out just sufficiently to clear the flying circle of the bolt ends, and the groovers would be set out a distance beyond the line of the planing cutters sufficient to cut to a depth of 12 mm. This distance would *not* be 12 mm, for (again in plan) the planing cutters would generate a particular size cutting circle, as would the tips of the groovers. The difference between the diameters of the two circles would be 24 mm, but as the cutter face lines are chords to the circles and not normals, the distance between the tips of the groovers and the tips of the planers would be more than 12 mm, measured along the cutter faces. Thus the difference in projection from the block face between the two cutters would be more than the difference between their two depths of cut. Cutting depth is never the same as projected depth, and allowance must be made for this when making and setting cutters. Some of the cutter projection is always lost. As the cutting angle changes due to greater or lesser projection, so the rate of loss of projection will change. The lower the cutting angle the lower will be the rate of loss of cutting depth. This means that when allowance is made for losses in a curved or moulded cutter profile, more allowance must be made as the profile nears the block face, for, as was stated earlier, the lower the projection, the greater the cutting angle. At no time (other than for grooves and rebates) will square block cutters be the same in profile as the moulding required. Figure 6.8 shows

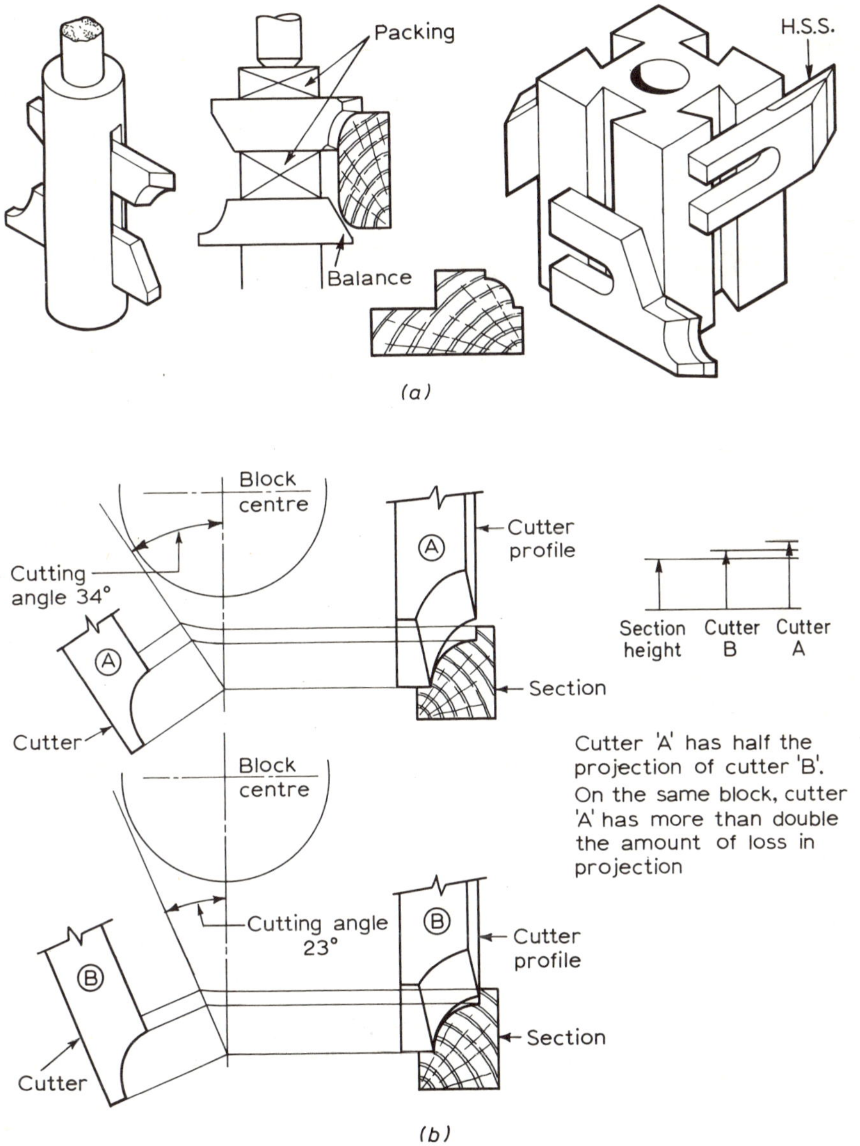

Figure 6.8 Cutter-block types and cutting angle principles. (a) *French head cutter with two cutters and* (b) *how cutters other than those in French heads lose profile depth when cutting*

the French head, square block and cutters, and a visual explanation of the basic principles that have been discussed.

Approach to profiles

The man on the shop floor must recognise the fact that cutter profiles vary from the moulding profiles they produce. He need not necessarily understand the geometrical principles. With experience he learns judgement, and a good cutter man is able to grind and set blocks from a section paper without any aid other than his judgement and a plywood setting strip. This judgement takes a long time to acquire and geometrical setting aids help to make accurate setting possible until this skilled judgement is attained.

The geometrical method used is to plot, in plan, points of projection at increasing distances from the block centre. If these points were fully extended they would form a series of concentric circles around the block, each circle representing a different projection of the cutter, with the smallest circle just large enough to clear the rotating bolt ends. The line of the block face is projected to cut across these circles, or the short arcs that are all that it is necessary to draw. A parallel line to this block face line is drawn, and this passes through the block centre. The short arcs should connect both of these lines, and each arc plots the path of travel of a point on the cutter profile, or the paths of separate cutters set at different projections. The distance between the arcs on the centre line will be regular and each space will represent a regular increase in cutting depth. The spaces between the arcs on the block face line (which is the cutter face line) decrease in size the further away from the block face they are, but they will always be greater than the measured distances of the spaces on the centre line.

The two series of distances are called *true units* and *projected units*, with the true units being on the centre line. The smaller the units used, the more accurate will be the aid in use, but the greater will be the possibility of inaccuracy when drawing. A reasonable unit would be 4 mm, which allows visual assessment of sub-units.

Having obtained the lines of true units and projected units, cutter profile plotting becomes easy. Salient points are marked on the section paper detail of the mould profile. Then, from the planing line (which will be the bolt end clearance line) the distance is measured to each of the premarked points. As each distance is measured, reference is made to the projected units on the block face line, and this second distance is marked on the section behind the original one. Thus if a square rebate 12 mm deep is shown on the section paper, reference to the projected scale will show, on the 12 mm line, a true distance of approximately 13·5 mm, and this will indicate the true setting position for the rebating cutter. With mouldings, a complete projected section is built up when all of the second series of marks are connected. The cutter(s) is/are made to this profile which also serves as a guide when setting.

The geometrical scales drawn can be used in several ways. The simple application described above can be extended further and a two-dimensional graph constructed with true units along the length of the block and projected units away from the block. Any moulding section could be picked out on this scale

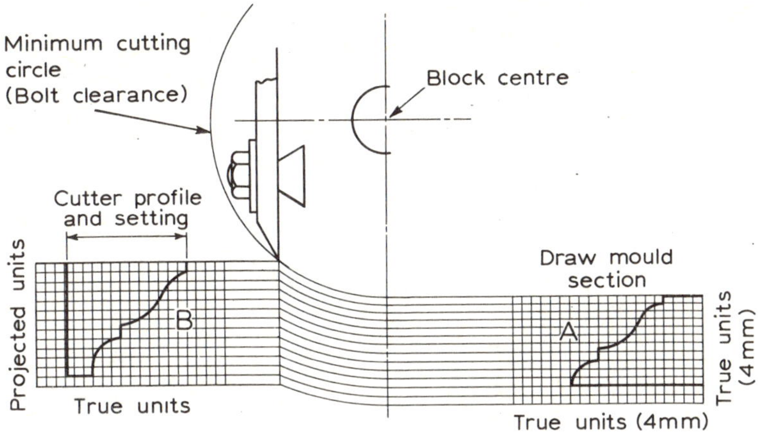

Scale A used to construct mould. Scale B used to project mould into a cutter profile and setting templet

Figure 6.9 Geometrical construction of a pair of projection and setting scales

and all points connected. The complete scale would then be used to grind and set the cutters. The complete scale and drawn section can be applied to a thin board. When used for setting this is called a *projection templet* or *setting templet*. Another application is to construct a 'projection rule' by marking projected units on one edge of a narrow plastic strip, and true units on the other. The mould section is measured with the true units, and points on the cutter face with the projected units. Figure 6.9 explains geometrical profile development.

Balance

All cutting tools must run in balance. This means that all forces acting on a block in movement must be balanced by opposite and equal forces. Equilibrium is obtained when all forces acting on a body are opposite and equal. A heavy cutter will not create equilibrium if it is positioned opposing a light cutter. The heavy cutter will act to distort the path of the rotating block.

A 'see-saw' form of balance, i.e. a beam, will rest at the horizontal when the downward acting force at each end is equal. A cutter block supported horizontally between needle centres will rest in any position provided that any cutters mounted on it are in pairs (for weight) and are set at equal projection, and that holding bolts are identical. The block is in 'static' balance, i.e. balance at rest. All forces acting to rotate the block clockwise are equal in sum to all forces acting to create anti-clockwise rotation.

Hence the first rule of balance is 'Cutters must be in balanced pairs and be set with equal projection at opposing sides of any cutter block', this achieves static balance.

When cutter blocks rotate another force is involved, i.e. that of centripetal force which is the force that must be present to stop the cutter flying off. A basic law of motion is that once in movement a solid body will endeavour to continue to move in the same straight line and will resist efforts to change its path. A cutter flying off at a tangent (a straight line from a circle) will fly in a straight line until the force of gravity overcomes its momentum. A cutter forcibly restrained from flying off still has the desire to do so, hence the forcible restraint. This restraint, exerted by the nut and washer, is the force required to 'bend' the cutter continually out of its chosen straight flight. This force is 'centripetal', and is the force that acts to the centre. The greater the velocity of the cutter the

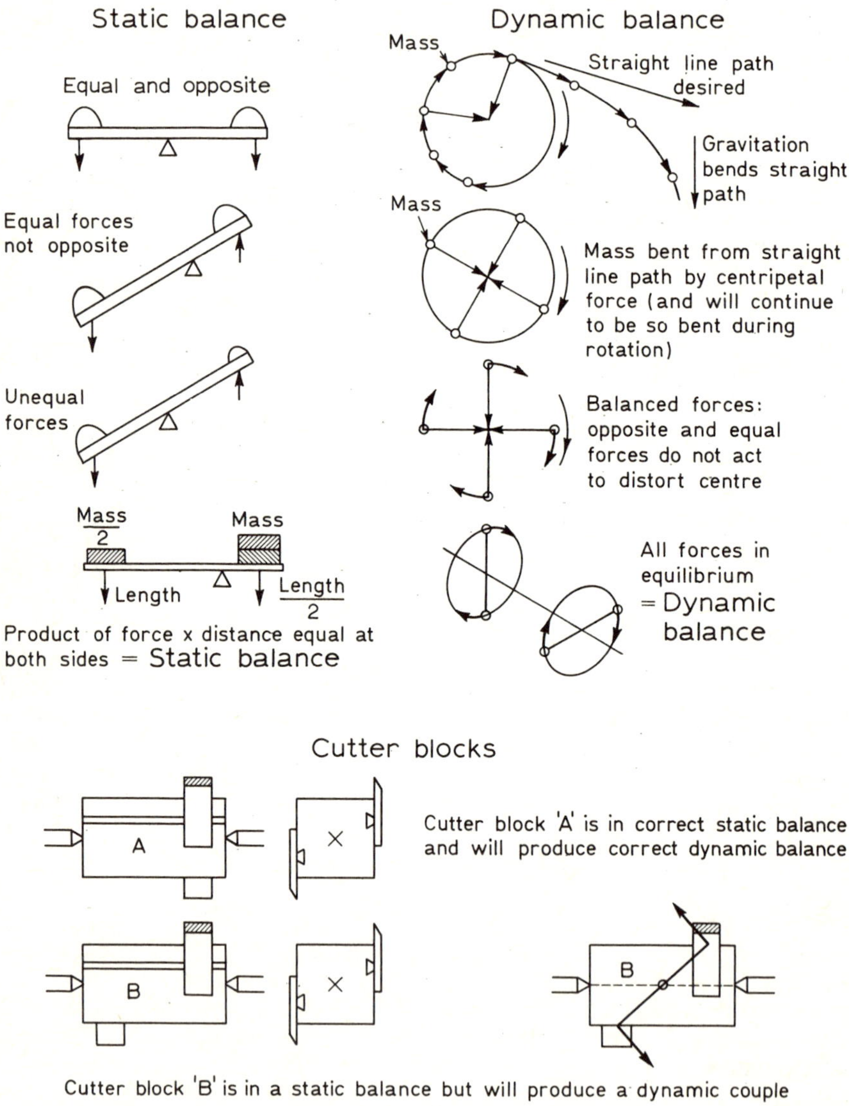

Figure 6.10 Principles of cutter balance

greater will be its desire to continue in a straight path and the greater will be the force necessary to restrain it. To be in balance, forces must be opposite and equal, and the opposite and equal force to centripetal force is centrifugal force, i.e. the force that acts away from the centre. As the inward pull of centripetal force increases at higher speeds, so does the centrifugal force.

As the revolutions of a cutter block increase so will the velocity of any cutter on it, and so will the 'bending' force of the pull to the centre. The increase in force is not proportional to the increase in velocity. Doubling the speed does more than double the force. Centripetal force increases as the square of the increase of velocity, which means that at double the velocity the force is squared, i.e. quadrupled. A three-fold increase in velocity causes the force developed to be increased nine-fold, i.e. the square of three.

As centrifugal force is always opposite and equal to centripetal force, and because it is easier to visualise an object flying away from a centre, most problems involving these forces are resolved in terms of centrifugal force. If centrifugal force increases as the square of the increase in velocity, it follows that two different masses (weight cutters) will have enormously different forces acting on them at the same velocity. A small weight doubled is still a small weight, but a large weight doubled is massive, and a small weight which increases by the square of any increase in velocity rapidly becomes massive.

Hence a few grams difference in weight between two cutters at rest becomes a difference in kilogram force when the velocity is increased to woodworking speeds, and the situation does not rest here. Not only must the outward pull of each cutter be equal in force, but it must also act at a point immediately opposite to its pair. Only when this is so will the block be in 'dynamic' (running) balance.

Hence the second rule of balance is 'Cutters must be set opposite and equal when set to a block'.

One other facet of balance requiring attention is the 'dynamic couple'.

If a block in static balance has one cutter of an opposing pair set towards one end of the block, and the other cutter set towards the other end, each cutter will exert the same leverage when at rest and the same centrifugal force when in motion. But the split pair of cutters will have their forces acting at opposite ends of the block, and the rotating block will endeavour to turn end to end about a point between the two cutters. This situation is known as a *dynamic couple*. The block will not attain dynamic balance. There must be no dynamic couples in a rotating block, all cutters must be in equal pairs and be set opposite and equal in projection.

PRACTICAL EFFECTS OF UNBALANCE

A discrepancy at low speed builds up to a massive discrepancy at higher speeds. This effects the work surface and creates undue wear of bearings. In extreme conditions vibrations can be sufficient to cause a cutter to fly off.

Blocks must run in balance. Often, a quarter turn or more of the block around the spindle may be sufficient to cure unbalance, especially if it is caused by a slight distortion of the spindle.

Figure 6.10 illustrates balancing principles. See Chapter 12 for calculations.

Other cutting tools

Cutter profile problems are common to all forms of cutting tools other than those used in French heads. No cutters other than French head cutters have the same profiles as the mould they produce. Balance problems are present with all forms of cutter blocks where loose cutters are fitted. Where cutters are solid profile cutters or blocks in themselves complete, or where cutters are standardised and merely fit into prepared slots, no balance problems arise.

SLOTTED COLLARS

Designed primarily for use with shaped work, slotted collars are simple in design and reasonably safe in use. They consist of two flat collars positioned one above each other on the plain spindle. The two facing surfaces have parallel grooves milled across them. One groove passes to each side of the spindle. Ground-steel stock cutters fit into the grooves in pairs and are held therein by the downward force of a large nut screwed down the top section of the spindle above the collars. One end of each cutter is shaped for its purpose and projects beyond the collar, unless, for straight edging, the two cutters are set flush with the collar edge.

Both cutters must be identical in width to ensure firm clamping. As there is no method by which one cutter can be adjusted vertically relevant to its pair, cutter making must be very accurate. As there are only two cutting edges available, collars will not remove such heavy cuts as will square cutter blocks. Balancing is simple if both cutters are ground closely to the same shape. Cutters are mild steel with alloy or h.s.s. facings.

For shaped work the edge of either top or bottom collar can be used as a running ring for the jig or templet, or for the previously machined edge of a curved workpiece that requires further moulding. To save wear on jigs and templets, collars can be fitted with ball-bearing rings which rotate freely and provide an easy surface for contact.

DISC-TYPE HEADS

There are many types of disc heads, the most common of which is the 'Whitehill'. These heads are steel discs with a centre hole for spindle and cutter clamps built into their periphery. Most have only two such clamps, but heads with four clamps are available. The difference between each make of head lies in the design of the clamps. Each method is patent to its manufacturer.

In all other respects the heads are identical. They have the same use and purpose. They carry small, solid h.s.s. cutters into the cut, supporting them very close to their cutting edges. Cutters are reasonably cheap and are easily made, and being clamped between faces, can be rotated through 360°, which makes it possible to use each edge of an approximately square cutter. Bevels are easily set. Each edge of a cutter can be shaped to a different profile, making cutters even cheaper.

Balancing is not difficult, for cutters are small compared with the mass of the block and minor variations in weight and size have little effect on dynamic balance. Only the cutting edge is unsupported and disc heads have a built-in safety element in that the cutter is supported right up to its edge, and the block itself, being narrow in thickness, follows the cutter right into the cut, thus obviating the need for long projections.

Disc heads can be run in pairs for tenoning or to carry cutters that are too wide (long) for one head, part of each cutter being held in each block. These heads are one of the most popular types, being easy to set and easy to use. They cut with a large cutting circle which gives good quality cutting. They are not suitable for small shapes or jig work unless a ring fence or roller collar is used.

THE SOLID PROFILE CUTTER

These are self contained, each being a complete cutter and head. No loose parts are fitted so no problems of setting or balancing arise. The cutting profile is built in by the maker, and balance is assured because they are symmetrical. Setting time is much reduced, but they are expensive in first cost and require precision grinding to maintain sharpness. There must be sufficient repetition required to justify their use. Most commonly they are used for rebates, bevels, nosings and grooves, where they are adjusted to size.

The cutter blank (other than for those to carry carbide tips) is made from solid h.s.s. or alloy steel. The disc, which can be cast or drop-forged to modify the metal structure, is turned on a lathe or end mill to the profile shape required, i.e. exactly opposite to the section required. Notches are then cut to form teeth and the head can be given two, three, four or six; for spindle working four are usual.

After toothing, the edge or periphery is finish ground accurately to the cutting shape required. As the disc rotates in cut, this shape is generated by the cutting edges formed by the teeth and no problems of loss of projection or cutter profile development is encountered. The metal behind each tooth edge will rub, as there is, as yet, no clearance. The relief is now ground in. This can be curved relief or straight relief, the former being geometrically the best. The relief grinding follows the edge profile for the length of the tooth back. Thus when the cutter is re-ground for sharpness with the original cutting angle maintained, the section cut will remain true to shape. Some side relief is given but only that which is sufficient to prevent rubbing when the cutter is sharp. The trend is to use pre-set and pre-formed heads as much as possible.

Grooving saws, standard bevels, rebaters, square-profile circular blocks, adjustable wabble saws, corner lock cutters, finger jointers and shaped edge jointers all advance production and reduce changeover and setting time.

The most recent innovation has been the introduction of circular heads using disposable tungsten-carbide tips. These are mainly for rebating or square edging where they are used to trim plastic laminate sheet back to its base of plywood or particle board. The tips, which are solid and of varying shape, have more than one cutting edge and the number is usually four. Each tip is positioned with one cutting edge presented to the job. When sufficient cutting has taken place to blunt this edge the tip is turned through 90° and a new edge is presented. When all four edges are dull the tip is discarded. This is a waste of carbide, but compared to the time wasted from production if the tips are ground, the loss is minimal. Grinding carbide requires time, good grinders, green grit and diamond wheels, all expensive items. The tips are all accurately ground to dimension to make turn-round easy and accurate, so, on balance, the discarded tip is more economic than it looks.

The use of tungsten-carbide-tipped saws and cutters is now so common that it has influenced the normal machinist's function of grinding and setting his own cutters because carbide cannot be ground on so casual a basis; it requires much equipment and skill.

Figure 6.11 shows a variety of block types.

Cutter grinding and setting

French head cutters are of high-carbon steel. They are cut-off and ground to shape freehand, with their shape true to the mould required. They are made sharp by swiss file and have their extreme cutting edge turned over by use of a burnisher.

Square cutter block cutters have facings of alloy or h.s.s. and are mostly ground freehand to a profile geometrically developed, plotted on a prepared scale called a *projection templet,* or are shaped by educated guesswork. Their profiles are not identical to the mould section required due to an inclined angle of approach to the wood.

Square block cutters must be used in balanced pairs and a comparative beam balance is used for this. This is a straight beam with a flat cradle at each end and which is pivoted in the centre on a knife edge. Cutters are positioned one on either cradle and are checked as they rest in several positions: slot in, slot out, face up, face down, until it is seen that the pair is either in, or out of, balance. The beam see-saws and has a fine line pointer that swings across a dead-centre line. Observation will show whether the pointer is swinging equally or is tending to swing more one way than the other, thus indicating that one cutter is heavier. By rotation of cutters, the heaviest point of the heavy cutter can be found, and then ground until balance is attained. Best places for removal of weight are the ends of cutter legs, the heel of the grinding bevel, and, sometimes, the inside of the bolt slot.

SLOTTED COLLARS

These are ground stock, faced with alloy or h.s.s. They must be in identical pairs and in balance. They lose profile depth in the same manner as square block cutters. Profiles are usually identical but can be slightly off-set up or down in parts to allow sharp corners to be cut. They are usually ground freehand, and setting only requires that they are tapped forward or back.

Spindle guards

Spindle guards are stereotyped. There are several standard and patent types in use. Most serve as pressure units as well as guards. The aim of a guard is to

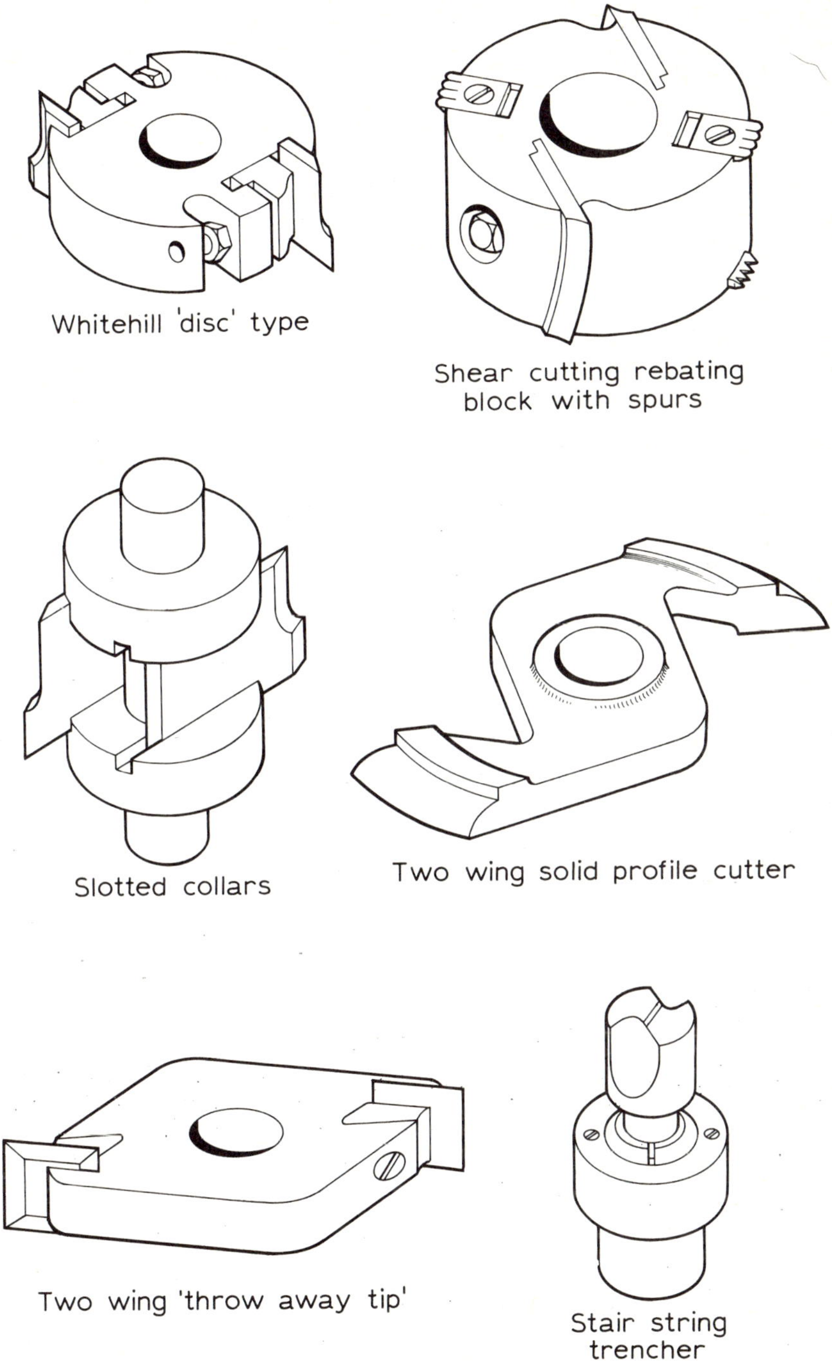

Figure 6.11 Further cutter block types

prevent contact between the hands of the operator and the cutting tools, to hold and steady the work to prevent it from being spoiled, to control flying parts of cutting waste, to restrain any flying or broken cutters and to protect the casual passer by. Some guards act as chip exhaust hoods.

The most common is the Shaw guard. In its simple form this is a pressure-pad unit which holds work down and obstructs operator contact with rotating cutters. In itself it does not satisfy the safety regulations. It will not contain flying parts nor prevent access to the rear of the spindle head. An additional bracket allows a horizontal side pressure pad to be added and this forms a channel with the fence through which long, straight mouldings can be fed.

Statutory protection is achieved when the Shaw is combined with the *cage* or *bonnet* which is an enclosing hood sited over and around the rotating tool, but with frontal access for wood. This guard can be vented directly into the chip extraction piping. In addition to the cage is an anti-snatch-back pressure shoe which is spring loaded to steady the work. This pressure shoe has a sharp edge which digs in if the work is thrown back by the cutters. The combined unit is called the *Yarwood.*

For shaped work where work, templet or jig is 'off the head' (no fences) or off from ball-bearing guide rings, a *Pen Farrell* is used. This is a steel pressure ring that surrounds the head and bears down on the work. It has a vertical flange to shield the cutting tool. Combined with a cage it gives statutory protection; without a cage it does not.

A 'ring fence' is often mistaken for a guard. These are guides used to allow shaped rails, etc. to be cut, and are designed to enable work to be firmly supported before it contacts cutters. There is a large centre hole that encompasses the cutter block. The rim of the fence is eccentric to this hole in that the guide ring is at its narrowest immediately on the line of cutting top dead centre. As the fence curves away from the cutting circle the guide ring gets wider until its overall diameter is greater than the cutting circle. Cutters project below the guide ring. Any templet or workpiece fed to the spindle is first pushed into contact with the ring out of the cutter orbit. Once supported, the workpiece is rotated around the outer periphery of the ring until the cutter is cutting at maximum depth. At this position the work is contacting the ring at its narrowest part, which is marked with a top dead centre line.

Power and speed

Wood is cut by power, and power is most effective if cutting is at speed. Spindle moulders run best at high speeds. The only limitation to free choice of speed is the size of the cutting tool. The faster the speed is, the better the job will be cut, but the cutter making and balancing must be more accurate.

French heads run from 4500 to 6000 rev/min, according to the size of the cutter and the diameter of the spindle. The larger the cutter and the longer its projection, the slower it must run (this is general to all cutting heads). Square cutter blocks can run from 3000 to 4000 rev/min, but if the block is low in height and very carefully set, it can run as fast as 6000 rev/min. Slotted collars, being almost always in balance and reasonably low in projection, can run up to

9000 rev/min, but a more common figure is 6000 to 7000 rev/min. Small high-speed spindles using special low-diameter heads can run up to 15 000 rev/min. Disc-type heads like the Whitehill run from 4500 to 6000 rev/min according to the diameter and the cutter size.

Solid profile cutters are always in balance. Unless they are very large the machine top speed is best, although those with the most teeth will cut as well at lower speeds as those with only two teeth at higher speeds.

The machine handbooks, if available, give information as to suitable speeds. The range of speeds chosen by the purchaser is usually calculated to suit a particular range of work and cutting heads. Standard spindle moulders have either two speeds or four, with reversing as an option. Shapers run at 7200 to 10 000 rev/min.

POWER

The power provided is again subject to the purchaser's choice. Standard spindle moulders use motors from 2238 to 3730 W, and the chain feed automatic requires 4476 W. The rotating table shapers have head power units that vary with machine size from 3730 W to the exceptionally heavy machine using 13·5 kW. The motor for table rotation is small, with an average of 1000 W.

7 Routing

Routing

Routing deserves the graphic description of 'to put to disorderly array'. Anyone entering a shop where a router is at work cannot avoid the sight of flying chips. The router works by the rooting and disturbing action of a small tool that penetrates a surface, roots around and scatters, in a similar manner to a pig and its snout. Some parts are left high, some are cut away and some are penetrated through.

The machine purpose

Broadly, router work can be divided into three groups, and the machine can be said 'to shape and edge-mould flat sheet material, recess into the face of material, perforate material,' and in general, 'to do as many other jobs as can be devised'. Figure 7.1 shows examples of each of these three groups together with additional cutting that cannot easily be classified.

MACHINE METHOD

Cutters, held in high-speed chucks (12 000–24 000 rev/min), are lowered into cutting position by the action of a foot pedal. The workpiece held in, or on, a shaped jig is then moved in a path that causes the cutter to cut to the pattern required. Control of cut pattern is achieved by means of a shaped templet attached to, or forming, the jig. The operator moves jig and job so as to maintain contact between the shaped templet and a steel guide pin that projects above the router table immediately in-line with the cutter. Generally, the pin diameter is the same as the cutter diameter, and this ensures that the job is reproduced to the same size as the templet. To make the job larger or smaller, different size pins can be used with the same cutter. Figure 7.2 illustrates these principles.

CUTTING SPEED

The exceptionally high rotational speed of a router cutter is made necessary by its small diameter. In Chapter 2 it was explained that a good optimum speed

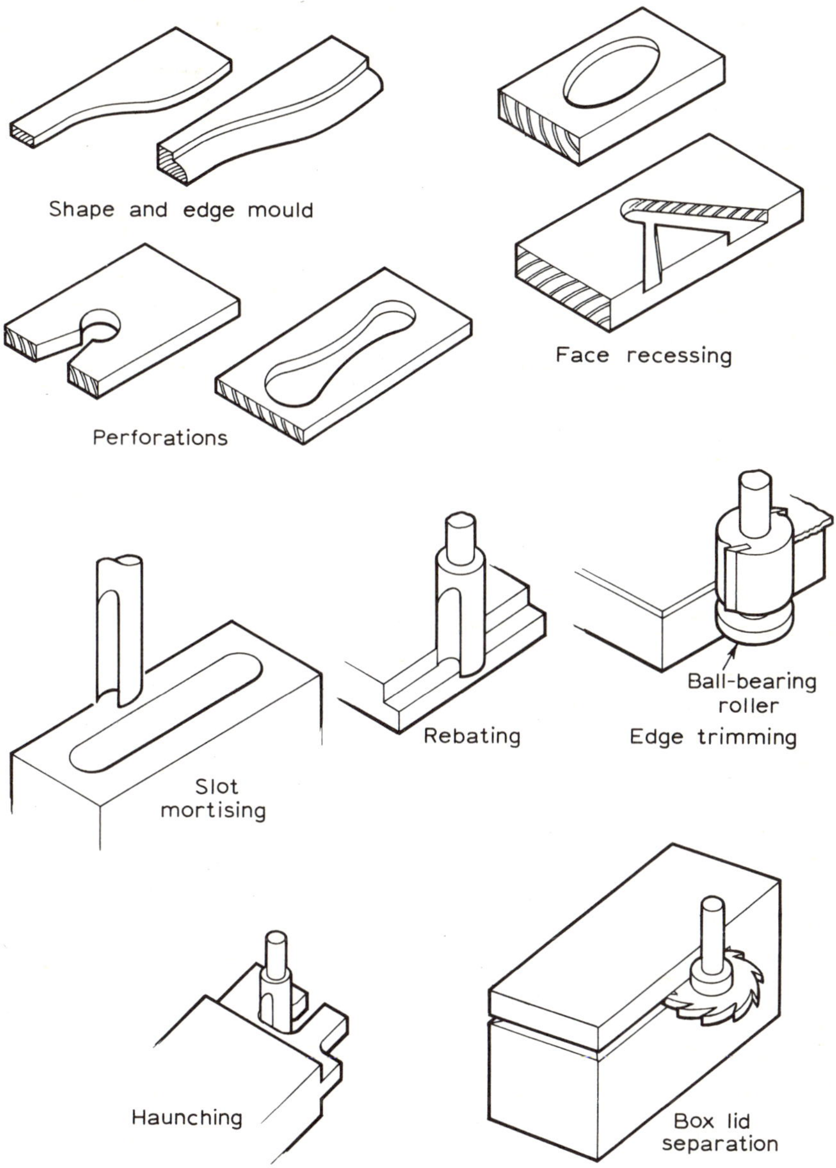

Figure 7.1 Router jobs

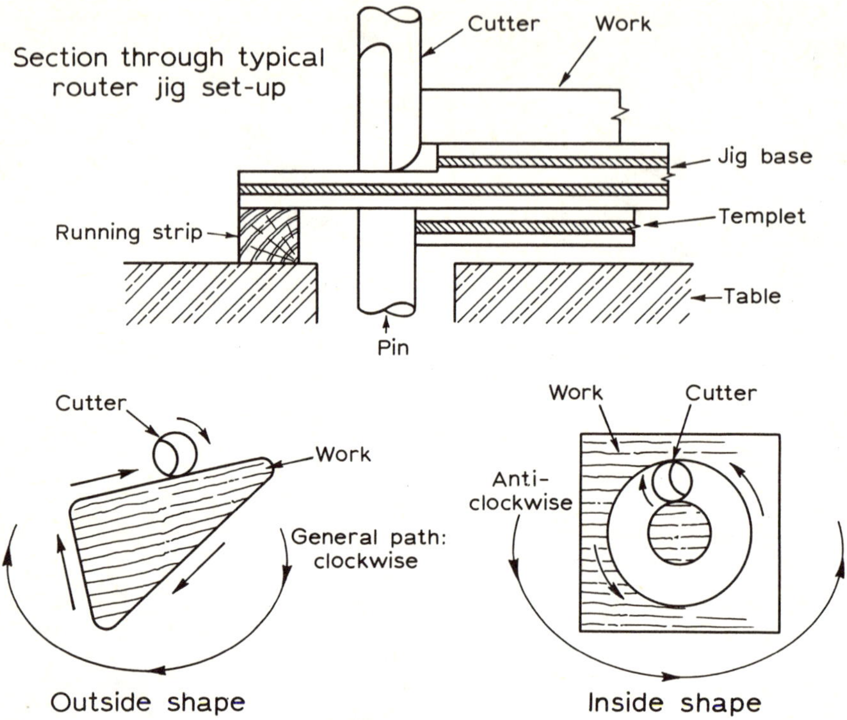

Figure 7.2 Routing method

for cutting wood is 1520 m/min. With cutters of 9 mm diameter on average, and that have a periphery of only 28·26 mm, high speeds of rotation are essential, and even at 24 000 rev/min the cutting speed of a 9 mm cutter is only 680 m/min, which is less than half of the ideal figure.

THE MACHINE

Routing machines are very similar in general appearance. There is a large swan-necked casting carrying, in slides, the cutter head and the work table, together with all the necessary stops and controls.

Work Table

The work table has a rise and fall movement controlled by a large handwheel through bevel gears and a lead screw. This rise and fall allows the factors of job and jig thicknesses to be related to the cutter and chuck length. The table will always have a hole immediately below the cutter through which the pin projects. Mostly, this hole is much over-size and is fitted with a filling ring. The hole allows large cutters to be brought down lower than the table surface.

There is normally no other movement to the table. However, to facilitate jobs where jigs are not necessary, or where work is to be carried in a straight line under the cutter, compound table movement can be built in.

The compound table is fitted with two sets of hand wheels and slides, one set at right angles to the other, and the whole table can be pivoted as one unit on the cutter centre line. Additional facilities available will be described under 'Special Machines'. Threaded holes are usual so that fences, clamps and other equipment can be bolted on.

The guide pin is held in a clamp below the table, and after initial positioning can be raised in either three or four stages, controlled by a small self-indexing hand lever. Each stage moves the pin approximately 4 mm. The pin, for normal use, must be set with its centre line on the cutter centre line. Free adjustment is not usually provided, the machine being centred on manufacture. After long

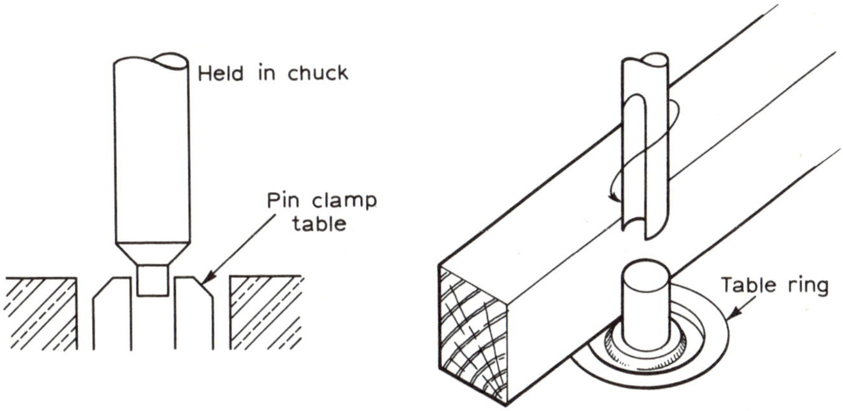

Figure 7.3 Method of centring the router table; for accurate work the pin must be centred on the cutter axis

use, or after the head has been moved (other than up or down) the table must be checked. For this a centre pin is provided that fits into the cutter collet at the top, and into the table pin hole at the bottom when the two holes are in line. The table-holding-bolt holes are drilled slightly over size and if the bolts are made slack there is sufficient movement available for table adjustment. The everyday shop-floor centre check is to use a square piece of wood that is pushed to the pin in several positions. A cutter of the same diameter as the pin is held in the machine chuck and pivoted by hand. It should brush the wood in all positions (Figure 7.3).

The Routing Head

The routing head is normally a built-in electric motor having the cutter chuck as an extension, through a shock absorbant coupling, of the rotor shaft. Two variations are the air turbine, where compressed air is fed to a turbine motor, and the belt drive where the motor is mounted at the rear and has a high pulley ratio to the drive shaft to attain high cutter speed. The air turbine may reach a speed of 30 000 rev/min.

The high speed of the cutter motor is obtained by the use of a frequency changer. Frequency changers accept electric current at standard mains frequency and change it to the higher frequency required to give high motor revolutions.

Motor principles

Frequency is the measure of the alternation of alternating current and is stated in terms of 'hertz' (Hz). At 50 Hz there are 50 complete cycles per second.

Induction motors of the 'squirrel-cage' type (most machine woodworking motors are these) can only run at speeds fixed by the frequency of the supply current and the number of wound poles.

Motors have a complex technology; all operate, however, because of the electromagnetic fields set up in and around any coil carrying a flow of electricity. A current flow is caused in any coil moving through, and cutting, the lines of force of a magnetic field. The greater the number of lines of force cut the greater will be the current flow and the greater will be the magnetic field or flux set up around the coil. Figure 7.4 illustrates the principles of the motor. The coil that is moved is the secondary coil and the current flowing in it when it is moved through the field of the primary coil is said to be *induced*, hence the name *induction motor*.

One magnetic field may be attracted to, or repelled from, another, depending on the respective polarities of the two fields. Like poles repel, unlike poles attract.

A motor is basically two coils, one of which moves round in the other. The outer motor coil called the *stator,* consists of the primary windings, and the moving part, called the *rotor,* carries the coil of the secondary windings. The squirrel-cage motor has a rotor laminated from thin iron sheets insulated and packed together. Around this ferrite core there is a cage of copper strips, each strip electrically shorted to a common copper ring, one at each end of the rotor. This is the 'cage' of the name. When alternating current is applied to the primary windings a moving flux, or magnetic field, is set up around the coil. This field rotates around the coils and thus around the stator. The field rotates because, in a three-phase system, the motor has three pairs of poles and each pair will 'peak' 120° later than the one in front of it, due to the phasing of the current through the conductor. The moving (rotating) magnetic field set up in the stator induces a secondary magnetic field in the rotor windings (the cage). The secondary field rotates and follows the primary rotating field. If this was all that happened the rotor would remain stationary although the two fields would fall into step, and, rotating together, become synchronous. However, the copper cage strips are embedded in the iron core of the rotor which reacts to the current and becomes magnetised, alternating from positive to negative in parts as dictated by the distribution of the copper strips (polarity).

Although the alternation of the current is very fast there must always be a time and space relationship; time for the iron to react, and space through which

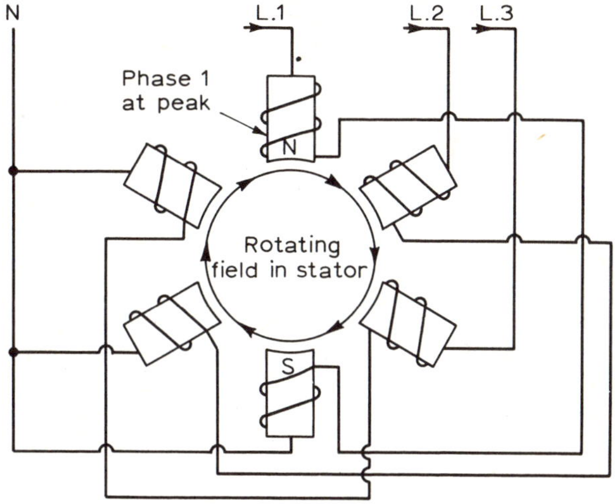

3-Phase 2-pole stator, 1 pair of poles to each phase

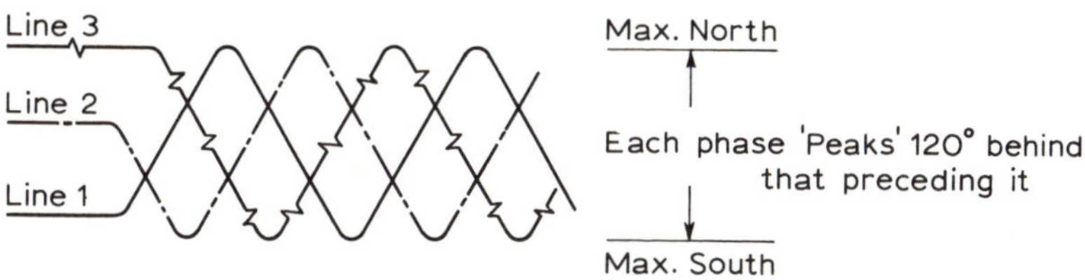

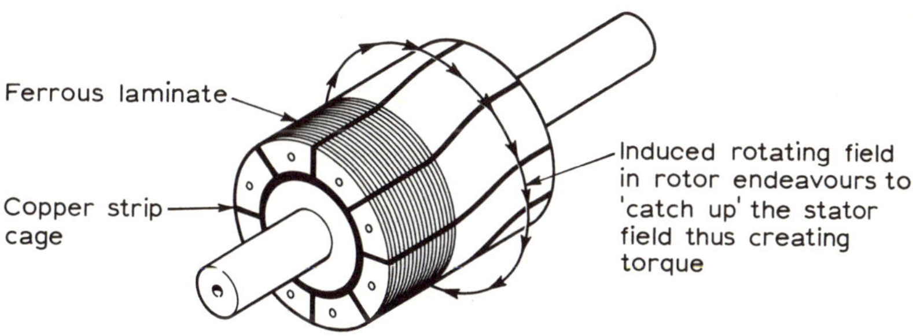

Squirrel cage rotor

Figure 7.4 Principles of the electric motor

the rotating field is moving. This time/space relationship causes the secondary field to lag behind the rotating primary field for two reasons. Firstly, the fraction of time necessary for the alternation of current to be induced in the rotor coils, and secondly, because the magnetic field set up in the iron core tends to act against the rotating field.

Other factors, such as bearing friction and windage, add to this time lag and the efforts of the secondary rotating field to become synchronous with the primary create the torque which rotates the rotor. The lag is called *slip* and is approximately 5% for squirrel-cage motors.

SPEED CONTROL

A squirrel-cage motor, with stator wound for two poles has a nominal speed of one revolution per cycle of frequency. Thus at 50 Hz there will be 50 × 60 rev/min, i.e. 3000. With the stator wound for four poles there will be only 0·5 rev/cycle and the motor nominal speed will be 1500 rev/min.

Speed can be varied by switching if there are enough wound poles in the stator, but as the basic speed with one pair of poles is only 3000 rev/min this would not give a sufficiently high speed for router work. Standard current at 50 Hz is fed into a frequency changer and given out at 400 Hz. At 400 Hz the rotating magnetic field speeds up with consequent increase in motor speed. At 1 rev/cycle this gives 400 × 60, i.e. a motor speed of 24 000 rev/min, minus slip.

The head motor will have a multi-pole winding system that will allow the selection of two speeds.

HEAD MECHANICS

The head unit contains bearings which can be oil-mist lubricated. A flow of light oil is drawn into the bearings by a felt pad. The centrifugal force of rotation then atomises it into an oil mist by turbulence. Externally to the head is a friction-brake press button and rotor lock, which enables the chuck to be held for cutter changing. This can be interconnected with the electrical control to cut off power when the brake and lock are applied. To bring the cutter into the work, the head is lowered; a foot pedal controls this. The Interwood range of routers are designed so that the head is lowered when the pedal is raised, whilst the Wadkin system operates to lower the head when the pedal is depressed. Wadkin make a lighter range of machines which have the pedal operating in the same way as the Interwood machines.

Both systems are efficient. With the pedal used to raise the head there must be a 'click-lock' to hold the head up until the next cut is required. A quick-acting hand lever enables the operator to cancel the hold of the spring plunger of the click-lock. Where the pedal is used to lower the head, the click-lock is used to hold it down during the cut.

Head movement upwards is by tension spring concealed in the main casting. In addition, the Wadkin heavy-duty router has a small rachet-adjusted pedal auxiliary to the main pedal, which is toe operated. The rachet permits progressive depth control for heavy, deep cuts. It also serves to hold down the main pedal when the cutter is cutting and can be used in this way to cut out the use of the click-lock. The head has a variable depth stop in the form of a quick-acting turret having four adjustable depth stops. This is pivoted, and located by a spring-loaded ball, which is turned by hand. Use of the depth stop enables multiple depths to be cut without changing the jig or the machine setting.

The Interwood router can be fitted with an automatic turret-depth stop with up to seven variations in depth. After pre-setting, the head indexes to the next depth of cut at each upward lift of the head. Some router heads may be pivoted through 90° and this facility greatly increases machine scope. Edge-on grooving and 'V'-cutting become possible using standard parallel cutters. Details of all these devices vary, but all serve the same purpose, i.e. that of increasing the machine potential and reducing operator fatigue.

A further feature of the head unit is the 'blower', a fan draught collected from the rotor and directed to the job close to point of cut. This blows chips away and directs them into any suction plant pick-up.

The range of power for routers varies from 1500 W to 6350 W.

CUTTER HOLDING

Cutters are mostly held in chucks or small cutter blocks. They can be made with shanks that fit directly into the machine spindle nose with adaptation for size by collets. The various systems are shown in Figure 7.5.

The internal bore of the head spindle is usually tapered upwards. Into this tapered hole a tapered cutter or chuck shank is pressed by the action of a locking ring that screws on the threaded outside of the spindle nose. The Interwood system has a short threaded length of shank on the chuck (or large cutter) and on this the locking ring is screwed. Further rotation screws the locking ring up the threaded spindle nose, thus pushing upwards the tapered chuck shank into the hole in the spindle nose where it is gripped by friction.

The pull of the chuck spanner locks the chuck into the machine. To release a chuck the taper grip is broken and then the locking ring is unscrewed from both the chuck and the spindle nose. The Wadkin system uses all parallel shanks to cutters and chucks. There is a tapered collet with internal, parallel bore to suit the shank required. There are three slots in the collet sides to enable it to close and grip the cutters. The shank is fitted into the collet and then the locking ring, which has a bevelled hole to suit the collet, is screwed up the threaded spindle nose. As the locking ring rides up it pushes the collet up into the tapered bore where it closes and grips the shank and is itself held by the locking ring.

Cutters and chucks

Cutters can be grouped into the following types:

1. Shell or spoon bits run in eccentric chucks, and which have single cutting edges.
2. Shell or spoon bits run in collets, and which have single cutting edges.

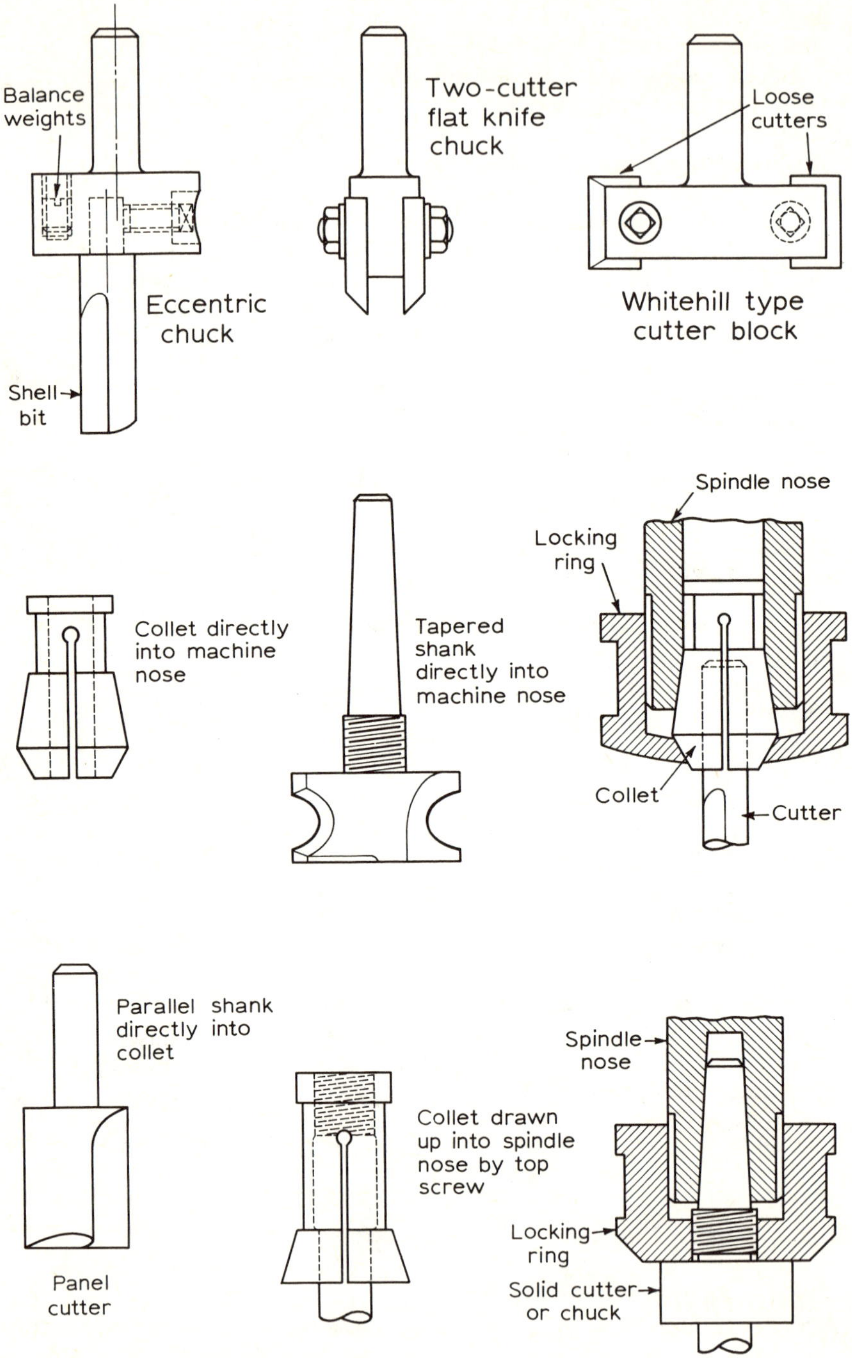

Figure 7.5 Methods of router cutter holding

3. Flat cutters held in chucks or small cutter blocks.
4. Double-edged panel bits; parallel.
5. Double-edged panel bits; solid profile.
6. Small, circular, solid-profile cutters that have a centre hole and that fit to small spindle-nosed arbors. These can be profiled to shapes, bevelled, square for grooving, or can be in the form of small slitting saws.

A range of router cutters is shown in Figure 7.6.

CUTTER CLEARANCE

The standard method of cutter clearance applies to most router cutters; the cutting edge leads and is ground away behind to give clearance, which must be sufficient not only when the cutter is tested in the chuck against a static piece of wood but also when the wood is being fed to a moving cutter. After a cutting edge cuts and moves along, the wood also moves on and in this way presents another portion to be cut by the following cutting edge. Edge clearance must be sufficient to allow this progression. Failing sufficient clearance, the wood rubs the cutter back and generates heat and smoke if persistently fed; a burnt cutter results. However, as diameters of router cutters are small, and maintenance of cutting diameter is desired, very little clearance is given. Operators must balance the need for fast, free cutting, against rapid loss of cutter diameter, and they sometimes grind greater clearances.

Cutters run in eccentric chucks are exceptional in woodworking. Out-of-centre rotation involves problems of balance and setting. However, with routers, an eccentric chuck system operates very well. The cutters, which are single-edge spoon bits, have no natural relief at all; they are in effect cylindrical pins having one flute to create a cutting edge. They are set into a hole off-centre to the chuck axis of rotation. If the cutter leading edge is set to its maximum radius of gyration, clearance is automatic as the rest of the cutter runs to a smaller radius. Rotation of the cutter in the chuck as it decreases in diameter when ground, maintains the original cutting diameter. Slight rotation of the cutter relative to the chuck enables the precise dimension control of the cut wood. Figure 7.7 shows the geometry of eccentric chucks and cutters.

Special machines

There are many variants designed around the standard router system. Designs fall into two groups: those that are similar in principle to the standard router, and those which operate with a moving head and static job. The first group vary from the standard router in that there can be two cutting heads, or they can have very complex compound tables and in some cases, automatic controls.

SEMI-AUTOMATIC

Using standard heads and controls, a range of machines is available that have compound tables that allow table movement in two straight lines at 90° to

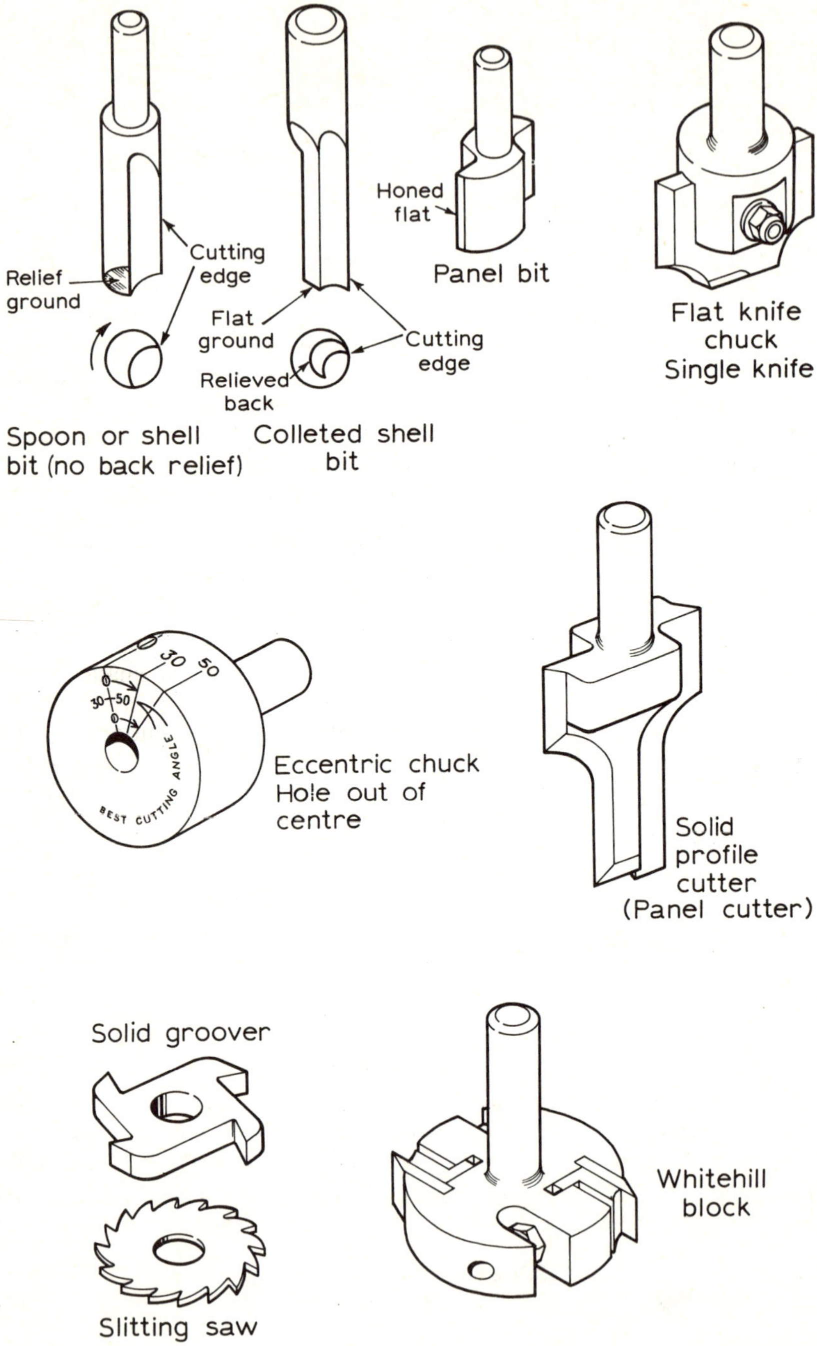

Figure 7.6 Router cutter types

each other. Work is clamped to the table and the table is moved by handwheel or lever between preset limit stops. An overhead templet can be used to avoid limit stops. This templet can either be incorporated into the machine design and elevated above the head unit where a projecting guide pin follows it, or it can be improvised to follow the cutter shank between the cutting end and chucks. The main use is for flush door and window openings.

A more complex arrangement has only the 'in' and 'out' movement controlled by straight slides and handwheel. Longitudinal movement of the table is controlled by a hand lever. At the back of the work table there is a ground channel which can be moved to suit angular requirements. A follower from the table fits into this, and for straight cuts parallel to the table front, the lever is used to pull the table along. Using two guides for the follower, irregular cuts can be made

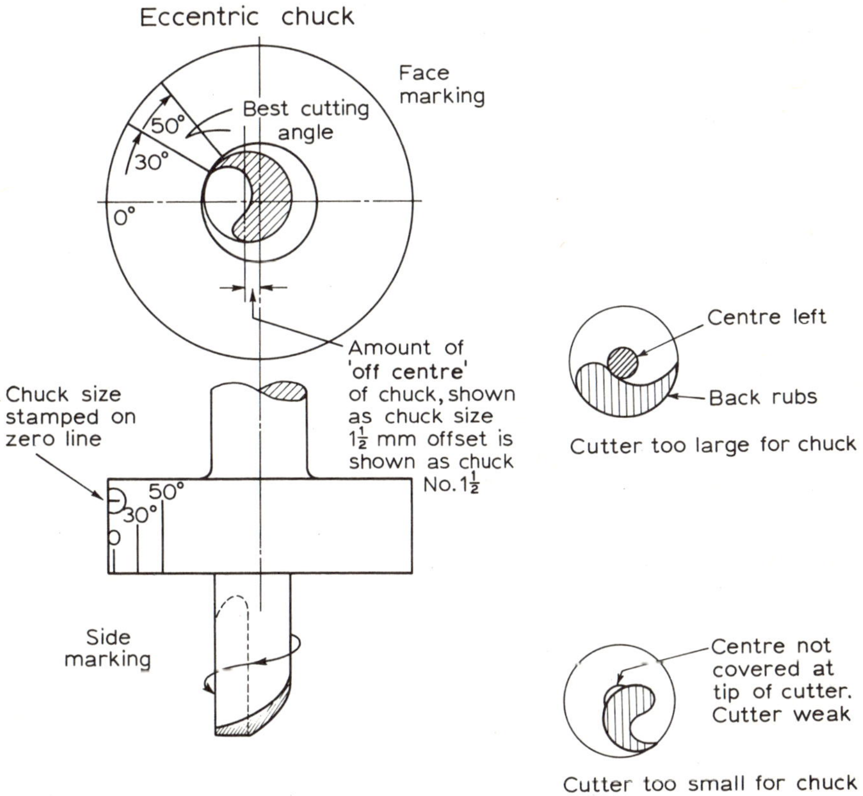

Cutting edges must be set between 30° and 50°. Less angle causes back of cutter to rub; greater angle creates loss of cutting 'hook'.

CUTTER SELECTION

Effective cutting diameter is equal to size of chuck added to diameter of cutter; subject to correct adjustment. Actual size cut will be slightly less than calculated because cutter is not set on maximum swing line (zero angle). Cutters must be 'right' for chuck. Cutters too large for chuck number will burn on back; centre does not clear. Cutters too small leave centre and tend to break.

Figure 7.7 Eccentric chuck details. Cutter and chuck must be matched for correct working, and the chucks balanced on a special balancing plate

without the templet. A shaped guide can be used for repetition contour routing.

A unique feature of this machine is the rotating stop control. A long steel tube is drilled with a large number of holes around its circumference and throughout its length. The holes are drilled at centres that give almost micrometer control of distance. By rotating the tube and selecting the correct hole for the spring plunger, table movement can be stopped at will to a preset position.

AUTOMATIC ROUTING

To obtain rectangular routing, the cutter head must move in two directions relative to the workpiece. Normally, the workpiece moves in two directions on the compound table slides. Most automatic routers are based on table movement for the lateral direction (in and out) and head movement on tracks for longitudinal (across front) movement. Several automatic machines are based on this.

Head and table traverse movements are usually based on pneumatics or hydraulics, or a combination of both, but some have electrical drive to one or both. Whatever the drive, the sequence consists of alternative head and table traverse until the job is finished. Presetting of head and table stops is all that is required; thereafter, all movement, including work clamping, is automatic.

A more complex machine is fully automatically controlled from a plug-board console, or punched card or tape reader. Once the console has received instruction, all work is fully automatic.

The above machines are essentially straight-line routers.

CONTOUR ROUTING

Imagine a work table free to move in either a longitudinal or lateral direction, then add double-acting pneumatic cylinders controlled by hydraulic check units to each direction. The table will now move in the direction of thrust of either cylinder if compressed air is fed. If both cylinders were fed together the table would move in a straight line at 45° to the front of the machine. Now feed the cylinders at varying intervals singly or together and the table will move in a path which is the resultant of the combination of movements. Furthermore, if the air feed to the cylinders is controlled to a plan, a predetermined contour could be followed. This is the basis of the automatic contour router.

An open-hole templet on the work table has its contour sensed by a stylus, movement of which controls a compound air-flow valve which feeds the table cylinders.

ELECTRONIC CONTROL

Contour cutting from the scanning action of an electric eye is another modern development. The machine has the standard cutting head mounted on a longitudinal track. The table is carried on lateral slides. A black line drawing is placed on the scanning table. Above the drawing there is a scanning head which contains an electric eye and beam projector. The principle is that if a light beam is projected, a certain amount will be reflected back when the beam reaches a surface. A photoelectric cell has the ability to read light values. A sensing screen is coated with a material that reacts to light in that its electrical conductivity is altered. Thus a light value directed to such a cell can be used to allow the flow of a minute current. By using amplifying relays this minute current can control the flow of working currents. The electric eye is therefore a switch, operating at the command of a light value.

With a sensing head the light beam reflects from the lighter part of the working drawing more than it does from the black lines. If the light source is surrounded by photoelectric cells each controlling a switch that independently commands the routing head to 'stop', 'go forward', 'go back', 'go right', 'go left', the power feeds to the rotating head and the work table will react accordingly, and a contoured path will be traversed. If, furthermore, command marks are strategically placed on the drawing, the head can be instructed to increase or decrease the cutting depth.

Replacing the drawing with a bandsawn silhouette pattern will have the same effective control, for the light will be reflected at a greater value from the silhouette than from the table below.

In short, it can be said that the traverse path will be moved to maintain the balance of light reflected to pairs of cells. If the light values get out of balance, the path is moved to correct this.

RADIAL-ARM ROUTER

All the machines previously described operate on the principle of the workpiece moving. The radial-arm router acts in the reverse way, i.e. the cutter moves around the work.

A beam pivoted at its centre and free to slide through its pivot has the ability to cover completely a given circular area with either of its ends. Figure 7.8 shows this and gives the design principles of the radial-arm router. This machine is suitable for larger panels, shaped chest tops, door top edges and table tops. Sheet-alloy sides and window openings for caravans are another application.

The fixed templet is followed by a roller mounted on, or above, the cutter shank and contact is maintained by hand pressure. For full outside contour cutting the work clamp pivots from side to side without releasing pressure. The job is cut in two sweeps. Stair-string routers operate on similar lines, with a templet above the string. Pivot and swing for the head is arranged through two arms; one pivoted at its inner end to a fixed centre, and the other pivoted elbow-wise from the outer end of the first.

Jigs

For standard router work, jigs are necessary. A jig can be defined as 'a work-holding box that contains all the necessary fitments to ensure adequate clamping, accurate location and correct repetition, and that at the same time ensures safe working'.

Jigs can be complex or they can be simple. Some jigs are simple templets pinned to the bottom of the workpiece and barely merit the description 'jig'.

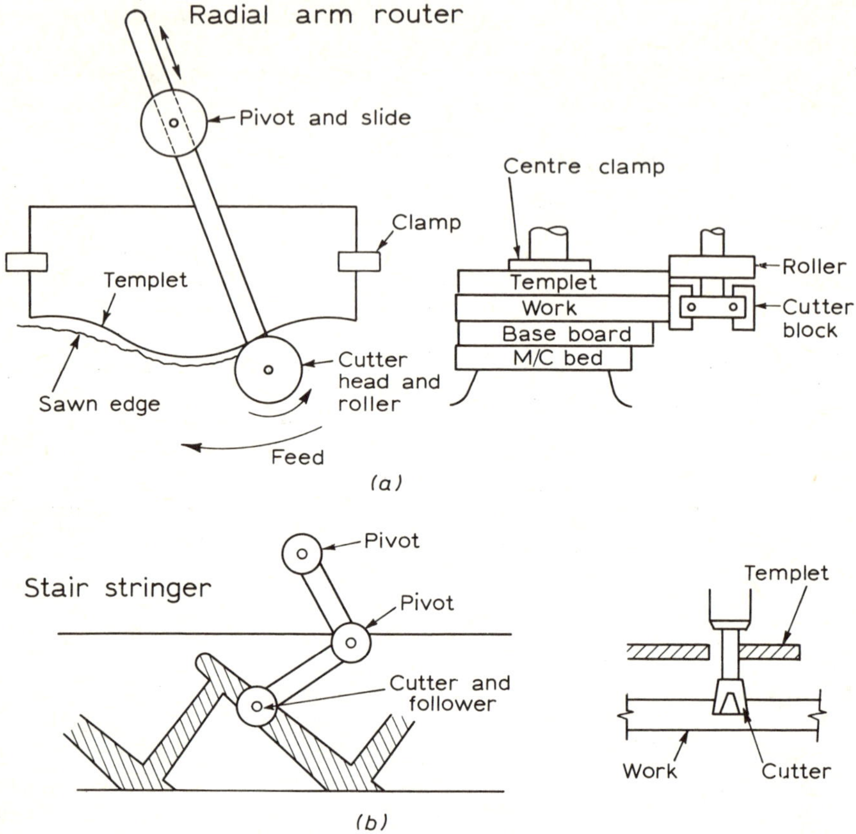

Figure 7.8 Routers. (a) *Radial arm and* (b) *stair*

However, the router is not at its best with such a tool. If the jobs are in production quantities, jigs are essential.

The jig components are:

The templet	To give correct dimension and shape.
Work locators	Fence strips and stops.
Work holders	Clamps, wedges, pins and cams.
The base	On which all components go.
Safeguards	Over-run and edge strips, and sometimes hand hold covers.
Hand holds	Means by which jig is safely held.

Figure 7.9 shows a typical jig and the above parts are all identified.

JIG DESIGN

Jigs must be safe in operation and accurate in repetition. This means that an area must be available for hand holds and that the mass of the jig is sufficient to absorb the shock of the cut. Work location must be accurate and work holding must not allow creep. Accuracy over both short and long periods is mainly a matter of using the best suited materials and maintaining cutters sharp.

Work must be presented to the cutters in the manner most suitable to avoid broken grain. Sequence cut should be preplanned in order that each cut removes break-out from previously cut corners, etc.

Figure 7.10 shows the various methods of work holding, but operators and jig makers often improvise methods other than those illustrated. A typical job and jig are shown in Figure 7.11, from the drawing to the work process.

Jig design and construction is learned on the job; experience is the only way. Of very great use are the two router handbooks produced by 'Wadkin' and

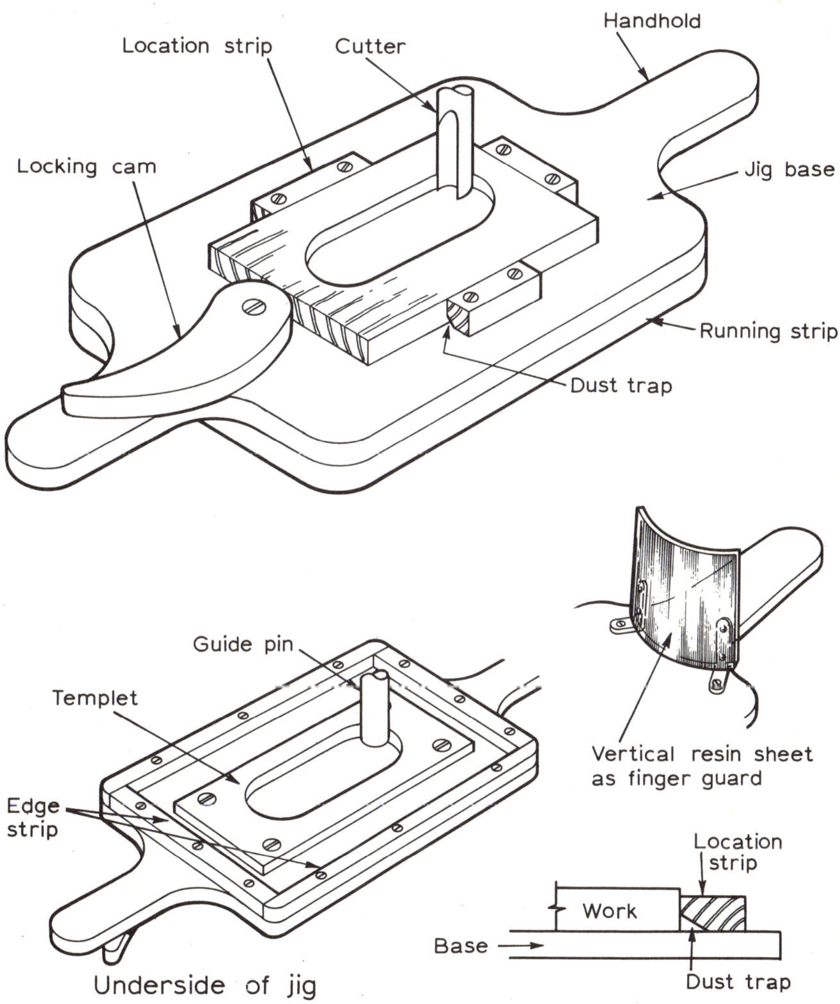

Figure 7.9 Jig construction

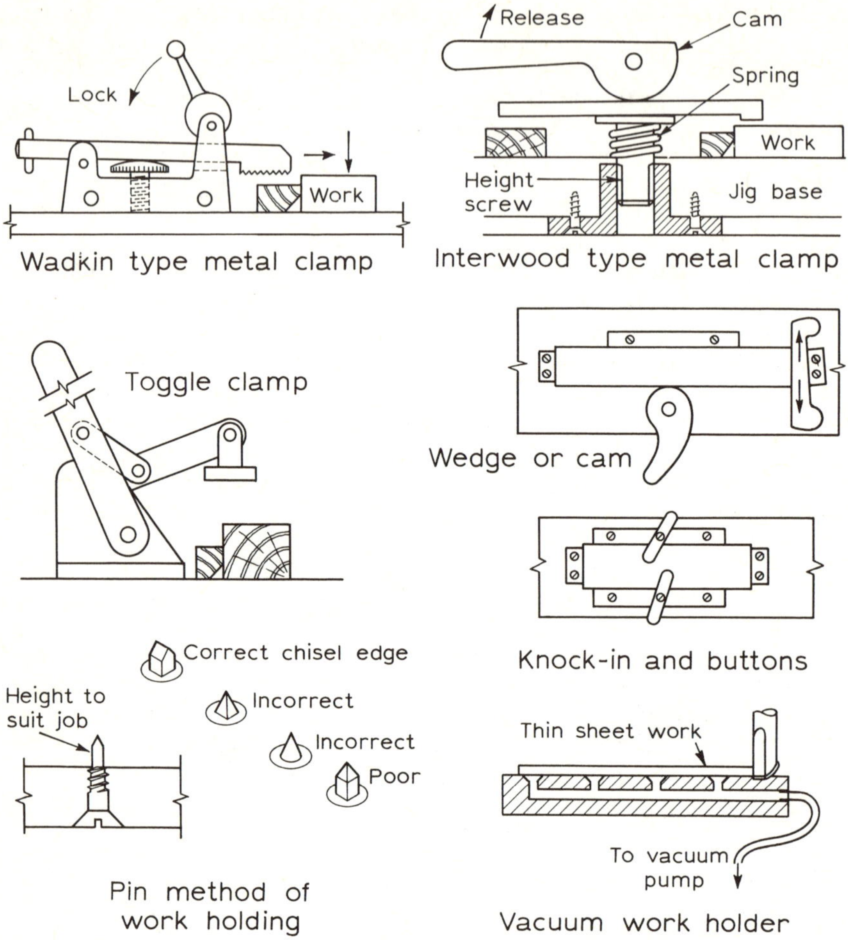

Figure 7.10 Common work holding methods

'Interwood'. These give a vast amount of information on jig design and router working methods.

Machine operation

Providing, as should always be the case, that the cutters and chucks are well balanced, the highest speed should be used. Only if large diameter cutters are in use is it necessary to run at the slower speed.

All cutters should be set with their cutting ends as close to the chuck or collet as possible. This cuts down vibration and lowers the risk of breakage. Cutting strokes should be continuous, for any dwell will burn the cutter and the wood. When penetrating, move the jig, for most cutters are not suitable for end boring.

To commence the cut, the jig should be pressed into the pin to prevent back-lash, and movement must be against the cutting edge. This means that for outside shapes the general movement of the jig must be clockwise, and anti-clockwise when the cutter is in a recess. The largest diameter cutter that will fit the shape should be used, for this gives cleaner cutting.

Guards

Router guarding is difficult because so often the jigs have projections above the job and cutters are short in length. The usual form of guard is the curved acrylic-resin screen riveted to a metal fitting. These can have either a sufficiently large diameter to surround the head working end completely, or they can be small and fit to the spindle nose. For maximum safety the larger type is best and usually insisted on by visiting Inspectors of Factories. It is static and shrouds the cutter whether the head is raised or lowered. The smaller type keep the operator's hands away from the cutter whilst it is cutting, but when the head is raised the guard goes up with it and leaves the cutting end of the cutter unprotected.

The problem of guards is that of bulkiness, which means that jigs must be large enough to give hand-room clear of the guard, and all fixtures on the jig must be low enough to clear. Another problem is that acrylic resin develops static electricity when it is vibrated and constantly struck by fast moving chips. This attracts more dust and makes the guard opaque. Polishing tends to increase the attraction. Anti-static creams are available but are not too efficient. Compressed air jets directed on the screen help to clear chips and dust.

The best safeguards for router work are the adequate planning of the jig and work method, together with the surrounding resin screen.

Tool maintenance

Router cutters can be ground free-hand, and very practised tool-room men do this. It maintains production and produces the work. The method puts up the cost of cutters, for grinding is more frequent and heavy. It also leads to the situation of 'just a little more off here', and 'lets take the back off just a little'. In practice it works, but only in the hands of an experienced and skilled man. With less skilled men it leads to 'bodged' work and burnt cutters.

Router grinding equipment is expensive as it is classed as precision machinery. Where large quantities of cutters have to be ground the appropriate equipment is essential. Smaller quantities can be ground on the router itself using various attachments, or the universal tool and cutter grinder can be used subject to the ancillary equipment being available.

Whatever the method, the aim is the same, that of restoration of the cutting edge to its original condition, which means that cutting angles are correct and there is sufficient clearance behind the cutting edge.

Grinding divides into approximately five basic jobs:

1. Inside of flutes of shell or spoon bits.
2. Lower ends of shell or spoon bits.
3. Cutting faces of double panel and profile cutters.
4. Lower ends of double panel and profile cutters.
5. Radial relief of panel cutters.

GRINDING PROCESSES

Wheels

Wheels used are standard aluminium oxide or white aluminium oxide. White wheels give best results for h.s.s. but tend to wear very quickly. For tungsten-carbide-tipped cutters, green grit wheels are used for roughing and diamond wheels for lapping. Wheel shapes vary from flat, with round faces, to taper and square cup wheels. Wheel size varies, according to equipment, from 35 to 100 mm diameter, but the large ones, although cutting more freely, are more difficult to fit to the cutter flutes.

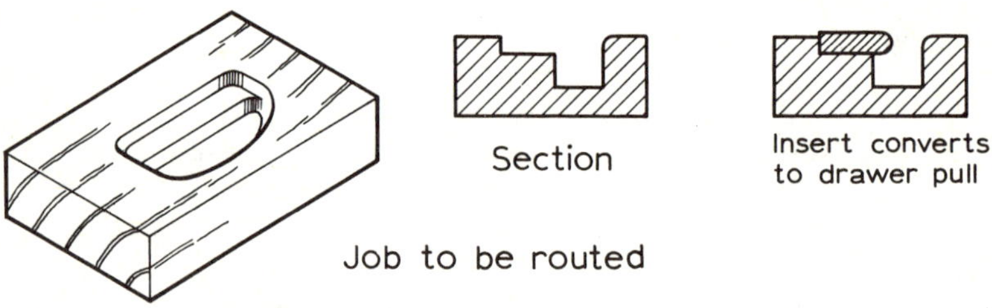

SELECT: largest pin and cutter that will give acceptable radius at corners. Cutter to have shaped profile.
METHOD: knock-in with buttons.
CUTTING METHOD: pin and cutter to same diameter. Jig to have double depth templet. Headstock to be set for two depths. Cut a shallow cut with pin down, then full depth with pin up.
RULE: with pin up the cutters must be down.

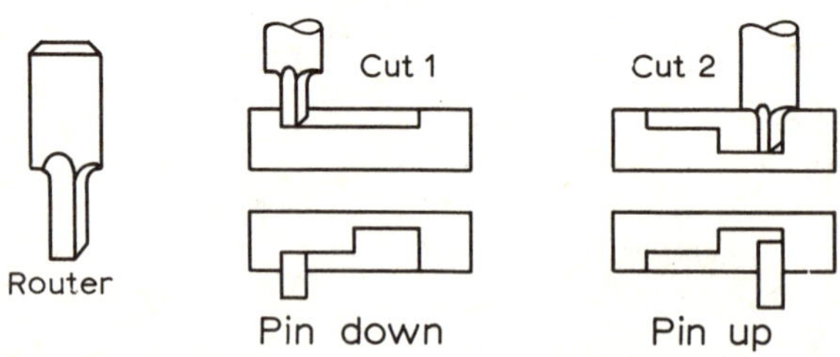

Draw templet shapes; cut accurately and finish to size.

Figure 7.11 Jig design and construction

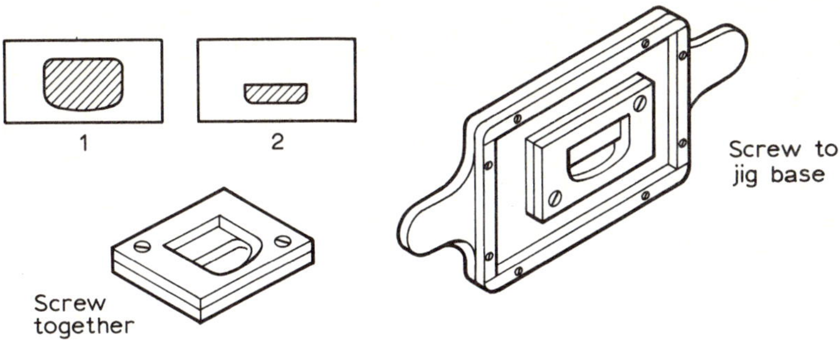

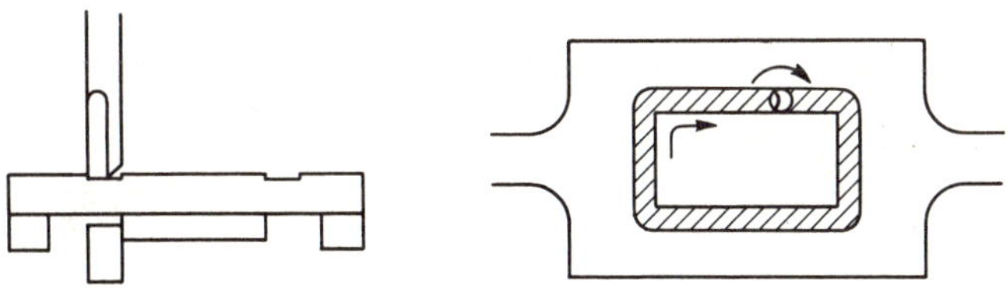

Set router with pin and cutter to cut 0·5 mm below jig surface. Carefully run round outside of templet, marking jig top.

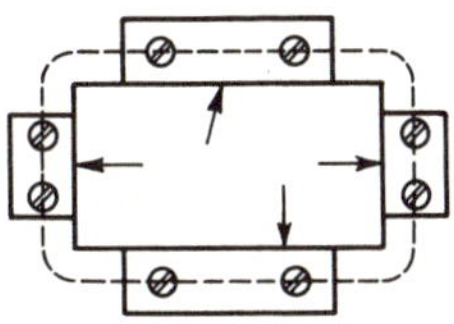

Use cutter track as location for location blocks. Check fit of test workpiece. Fit buttons.

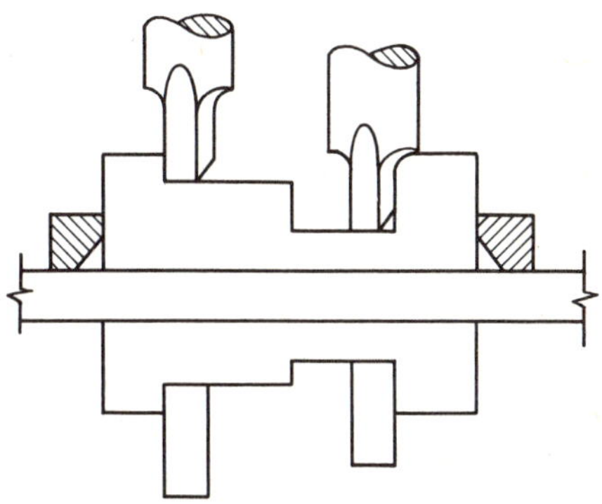

Fit pin to templet cut-out, setting just low enough to clear top templet. Set head stop for cutter to cut shallow part of job. Rotate pin adjuster to raise pin to deep part of job. Set second head stop and cutter depth.

Set machine guard. Make trial cut with pin down, cutter at first stop. Check shape. Reset pin and head depth. Check second cut. If long runs are needed use resin sheet as first testpiece; use parallel cutter, and then use testpiece as bottom templet.

Figure 7.11 (continued)

Shell Bits for Eccentric Chucks

Only the inner face is ground. The flute must be maintained with the cutting edge not too thin or too thick. New cutters form a guide. The outer face is polish ground by the makers and is rarely ground in the workshop. A light and delicate touch with a slipstone will easily remove any burr set up, but the slightest inward dubbing will cause the cutter to burn.

The wheel used is a flat round-nosed one. By holding the cutter angled across the wheel face, most of the flute can be ground. The wheel should run into the cutting edge. If the cutter is held parallel to the wheel, it must fit the flute and there is a risk that the cutting edge will be turned too far and lose its cutting angle. Figure 7.12 illustrates this. Cutter ends are ground free-hand keeping the bottom of the cutter flat and square to the sides. Enough relief is given to prevent rubbing.

Colleted Shell Bits

These have radial relief incorporated. The flute is treated the same as the eccentric chuck cutters. Where it is found necessary, the leading edge can be backed off with a fine, flat oilstone. This will produce a narrow flat land behind the leading

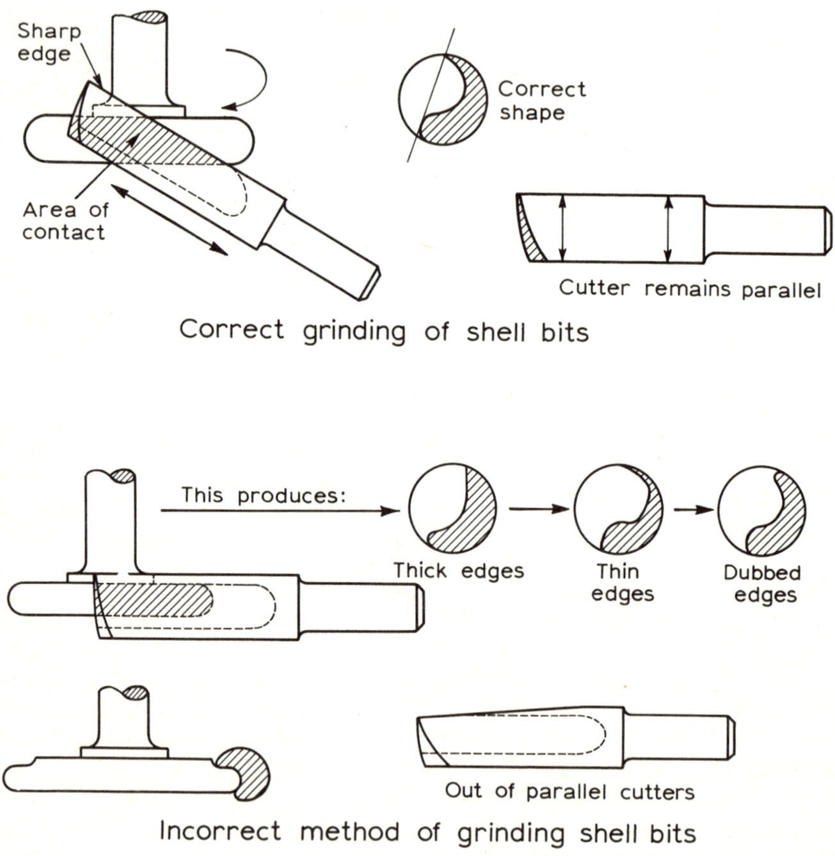

Figure 7.12 Free-hand grinding of shell bits

edge. Woolly timbers require more cutter clearance than more brittle timbers. For thermoplastic sheet (acrylic resins) the cutter must be sacrificed for the job and a much greater degree of radial relief is incorporated. The ends of these cutters are ground with a flat finish end which is angled upwards to give clearance.

PANEL BITS

These have two edges and mostly have square flutes and flat faces, although smaller sizes can have rounded flutes. Round flute cutters are mainly for non-ferrous metal routing. Panel bits are ground on the inner face of the cutting edge as shown in Figure 7.13. The grinding process is identical for parallel or profiled

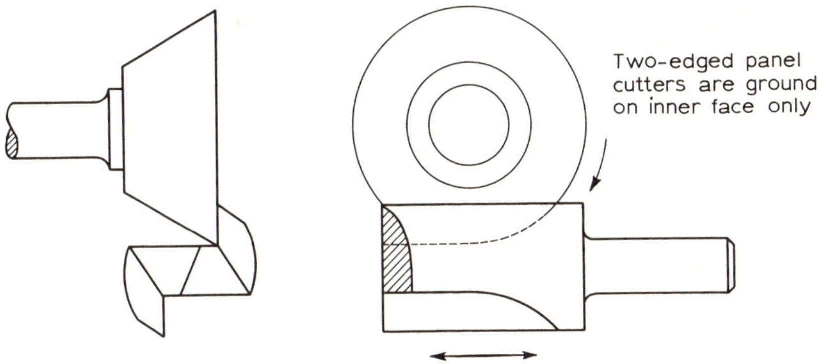

Figure 7.13 Panel cutters ground on the inner faces only

panel cutters. Where a poorly maintained cutter has to be reground it is best to check its angles against the most similar new cutter available.

The grinding attachment will have several movements: it will pivot from the horizontal; it will pivot in plan; it will index to suit a range of cutters. Simple attachments index for only two places, whereas more complex ones index for 2, 3, 4, 6, 8 and 12. The attachment will slide to and fro beneath the wheel, and it will have feed adjustment both inwards to the wheel and rotational round its shaft so that the cutter can be rotated towards the wheel.

The pivoting movements allow the cutter to be set at the correct angle. Shear-cutting cutters require tilting and pivoting. Indexing will permit multi-point cutters to have each cutting edge presented to the wheel in an identical manner. Slide, by push-pull or handwheel, enables each grinding cut to take place. Inward feed of the attachment facilitates correct angle setting. The rotational feed is of most importance, it enables the correct angle of hook to be maintained. If the in-feed of the cross slide is used to increase the cut of the wheel, the cutting angle of the cutter will change. Where a profiled cutter is concerned the profile cut on the wood changes as the cutting angle changes. Figure 7.14 shows the geometry of this. When both faces have been ground the ends are set to the wheei.

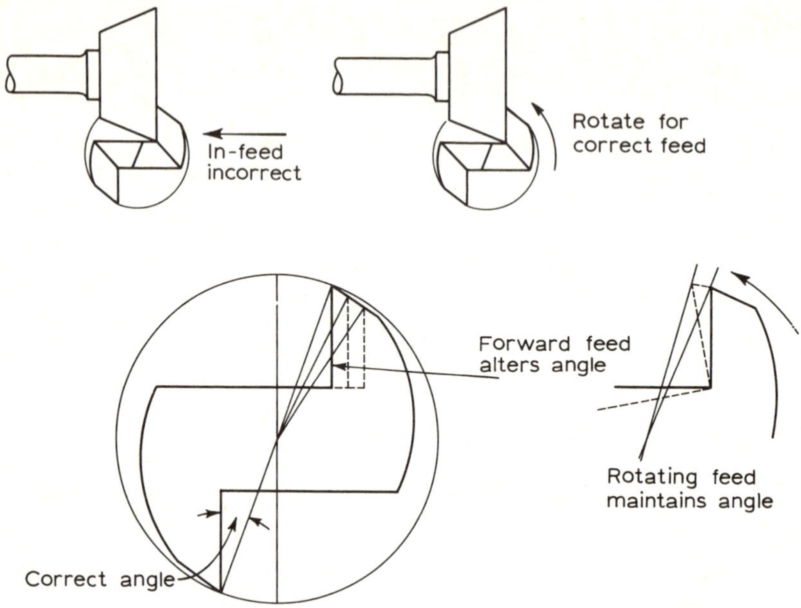

Figure 7.14 Maintenance of hook. The cutters must rotate into the wheel to maintain the correct angle and profile if shaped

Grinding process

The cutter is positioned visually. First contact is checked against other faces of the cutter. When the angles and contacts are correct, light cuts are made. Final polishing cuts are made to all faces without in-feed. As far as possible, the wheel should run into the cutting edge. With carbide tips this is essential, for the edges being brittle are easily knocked away.

After grinding (other than eccentric chuck cutters) a flat is slipstoned behind the cutting edge. This brings the edge to sharpness and ensures clearance. The stone must be held flat and must leave the cutting edge at the highest part. Small grooves and corner-lock cutters are ground on the back and not on the cutter faces. With these, the cutter width is of most importance. Face grinding would soon reduce their profile width because of their side relief.

FAULTS

Cutters that burn have poor clearance. A woolly, scraping action indicates insufficient angle of hook. Ridges left on the side edges of deep work cut in several cuts indicate that the cutter is tapered-in towards its shank. Eccentric chuck cutters that are run in a chuck of too low number leave a raised centre of cut. If their ends are not ground flat they leave a ridged centre.

Calculations

The optimum cutting speed for wood is 1520 m/min. The peripheral speed S_P is given by the equation

$$S_P = \frac{\pi DR}{1000}$$

Where S_P = Peripheral speed (m/min)
$\pi = 3{\cdot}14$
D = Diameter of cutter (mm)
R = No. of revolutions per minute (rev/min)

Example 7.1

Compare the peripheral speed of a circular cutter block of 100 mm diameter that runs at 5000 rev/min, with that of a 20 mm diameter router cutter running at 24 000 rev/min.
Cutter block

$$S_P = \frac{\pi DR}{1000} = \frac{3{\cdot}14 \times 100 \times 5000}{1000} = 1570 \text{ m/min}$$

Router cutter

$$S_P = \frac{\pi DR}{1000} = \frac{3{\cdot}14 \times 20 \times 24\,000}{1000} = 1507 \text{ m/min}$$

Comparison shows that small diameter router cutters reach almost the same cutting speed as that of a larger, slower block.

Power

Routers use motor heads whose powers vary from 2 h.p. to 8·5 h.p. As the speed increases, the horsepower output increases, and a 2 h.p. motor at 18 000 rev/min will give 3 h.p. at 24 000 rev/min. Horsepower is the accepted Imperial unit, but the S.I. unit is the watt (W), and 1 h.p. = 745·7 W. Hence a standard router motor for a medium machine with a 5 h.p. output would be rated as

$$5 \times 745{\cdot}7 = 3728{\cdot}5 \text{ W}$$

Speed

Squirrel-cage induction motors run approximately at synchronous speed equal to the rotation of the magnetic flux. For two-pole winding at 50 Hz this is a nominal 3000 rev/min; with four-pole winding it is 1500 rev/min. A change in either the number of pole pairs or the current frequency will change the motor speed. At 50 Hz there is 1 rev/cycle with the two-pole winding.

Thus motor speed will be:

$$50 \times 60 = 3000 \text{ rev/min}$$

At 400 Hz (from a frequency changer) the motor speed will be:

$$400 \times 60 = 24\,000 \text{ rev/min}$$

Due to 'slip' windage and friction the actual revolutions of a 50 Hz two-pole motor will be closer to 2850 rev/min.

8 The Four-Sided Moulder and Planer

The machine

The four-sider has many names that indicate the number of cutter blocks and its use. The most common is the *moulder* or *four-sider*. Older men still call the machine by its number of blocks, i.e. the four-cutter, five-cutter or six-cutter. Whatever the name, its job remains to plane and mould timber sections on all surfaces at one continuous pass through the machine. The machine essentials are the cutter blocks, the feed system and the pressure system.

The skills needed for fullest exploitation are the skill of assessment, whereby the best method of cutter arrangement is quickly seen and planned; the skill of cutter grinding, remembering that no shaped cutter fitted to a rotating block will produce its own shape, and so allowances must be made; the manual dexterity of setting cutters to the blocks; the mechanical dexterity that facilitates quick adjustment of all of the machine moving parts; the judgement that permits optimum pressure setting throughout the machine after a study of the work to be run; and finally, the skill of all machine woodworkers in the visual assessment of the alternative merits of each piece of wood and the best manner in which to feed it to the machine.

THE CUTTER HEADS

Cutter blocks either carry cutters or are themselves solid cutters. The shaft to which each block is fitted is called a *head* of the machine. Each head has a name according to its position through the machine. With four heads the arrangement is standard. The first head acts on the bottom of the workpiece, and is called the *bottom* head. The second head acts on the inside or fence side of the machine and is called the *fence* head. The third head acts on the outside of the work and brings it to width. This head is called the *outside* head. The last head, the *top* head runs above the work and reduces it to parallel thickness.

Where five heads are available the extra head is horizontal and can be added as a second top head or second bottom. When fitted as a second top it is placed immediately after the bottom head, and when fitted as a second bottom it is fitted at the rear of the machine.

With six heads the bottom heads are first and last with top heads between. The side heads are between the two top heads, with the fence or inner side head

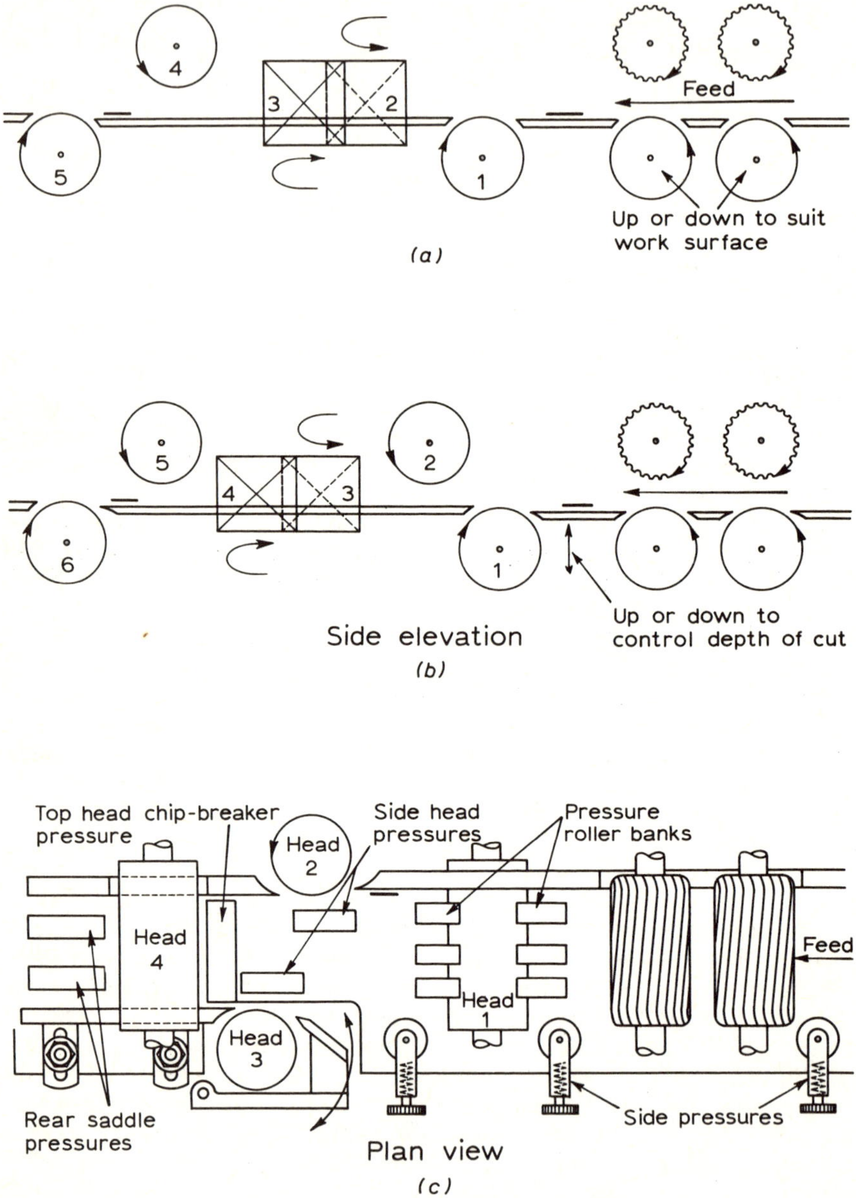

Figure 8.1 *Machine layout*. (a) *Conventional layout for a five-head machine*, (b) *layout for a six-head machine and* (c) *plan view of four-head machine*

always before the outer. Figure 8.1 shows the various arrangements of heads and a general outline of the machine.

ALTERNATIVE PLANS

As each new moulder comes on the market many variables are introduced. There are models with five heads, one of which can be adjusted either above or below work to act as an extra top or bottom head. Others have extra feed rolls mid-way through the machine and at the rear. This feeds each piece fully through and does not require that one piece is pushed through by that following. More full control is possible and less time is wasted at change over.

Other introductions have the feed roll pressure controlled by pneumatics, and push-button control allows increased or decreased pressure. Coupled with electro-micro switches, compressed air is used to sense oversize pieces and to regulate feed roll setting accordingly.

Standard moulders do not plane material out-of-twist. 'Stock' is brought to section and made parallel in width and thickness, but if twisted or bowed it remains in that condition and is rolled out-of-twist as it passes through, only to spring back as the pressure is released. Additional cutter heads are now fitted at the machine front, and these, together with a rolling slat bed and a form of finger feed unit, make the bottom face flat prior to entry to the moulder proper. Sometimes an extra side head is fitted to the fence side edge. This type of unit is in essence a power-fed surface planer.

THE CUTTER BLOCKS

These are standard and are basically identical with those described in Chapter 6. Cutting principles are the same, as is the use of projection templets.

For fast planing, multi-knife circular blocks are used carrying up to 10 cutters. Many experiments are being carried out to lessen the volume of noise to which operators are exposed. The two most promising have the cutters wound round the block periphery in a helical path. One has thin full-length cutters, and the other has a series of carbide tips at intervals, all dressed to cut in the same track.

The principle is that of reduction of impact vibration due to intermittent impacts of spaced cutters across a board width. Much of the total noise is caused by cutting air, as removal of fences, chip breakers and guards will show. This part of the total noise cannot be successfully eliminated; it can only be contained and insulated by elaborate screening and use of ear protectors.

Apart from circular and square blocks, the current trend is towards the use of solid profile cutters and this is aided by standardisation of joinery sections.

Block Mounting

Cutter blocks can be removed. Smaller moulders have no outside bearings for the horizontal heads, and the blocks simply slide on. Larger machines have outside bearings and these swing away clear of the shaft for block changing.

To utilise this facility, self-centring collars and collets must be used, for a simple interchange of block would not be sufficiently accurate. Self-centring equipment allows the block to be accurately registered on the machine, and also for setting on setting-stands, and grinding with cutters preset. Self-centring devices are based on tapered collets that can be split to enable them to be pulled into the block bore and thus squeezed around the block shaft. This locks the block on the shaft and registers for accurate centre. When end collars are used the collets are not necessarily split, but their bore must be accurate to the shaft. Examples of moulder blocks and self-centring devices are shown in Figure 8.2.

THE FEED WORKS

These consist of two pairs of nip rolls, one pair of which is below the work and the other above. The top pair is usually serrated or fluted to ensure full grip. Plain rolls at the bottom allow high feeding pressure without injury to the lower face, which may show if only light cuts are being made.

Top roll flutes can be angled around the roll perimeter and this causes work to ride into the fence. The top rolls are spring-loaded and have handwheel control for position. The bottom rolls are set slightly above the in-feed bed. All rolls are power driven by chain and/or gear train, from a gearbox. The prime mover is usually a two-speed motor to give two ranges of feed speeds. Some machines have a multi-pole motor which gives speed variation by switching rather than by gear change.

BEDS AND FENCES

Workpieces are fed through by the feed rolls. They move in a straight line because they are held up against a fence, which is in two parts, one before and one after the fence side head. The machine bed supports the lower face. The bed itself is complex.

Before the first bottom head is a movable platen which can be raised or lowered to control the depth of cut, and which can be moved towards or away from the block periphery so that at all times it is as close to the cut as possible. This prevents break-out. Behind the bottom head is the main bed which may be in sections, one each for the side heads and one for the top head. The principle again, is that the work shall be supported as close to the point of cut as possible. The top head bed may have a lead, white metal or wood insert to permit the cutters to break through. If a second bottom head is fitted the block is adjusted to suit the cut required and then the following bed plate is set to the line of cut.

Fences

To guide the workpieces accurately there is the side fence. This commences on the forward feeding bed before the feed rolls, and continues through until it

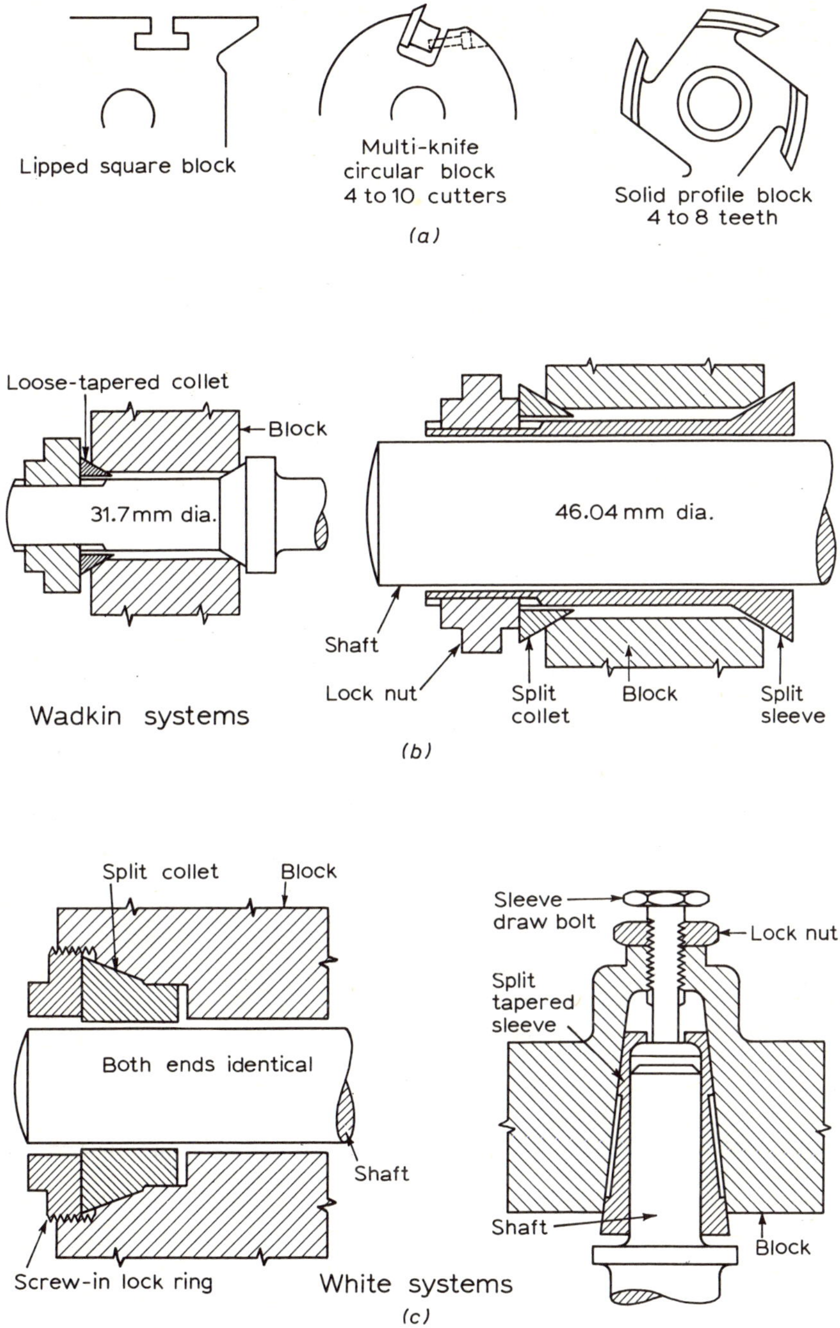

Figure 8.2 *Moulder blocks and self-centring devices.* (a) *three types of block,* (b) *and* (c) *methods of self-centring based on cone collets, which are sometimes combined with split sleeves*

reaches the cutting circle of the fence head. To enable cut control this fence will move forward or backward across the width of the machine to a limit of about 8 mm. Manufacturers incorporate alignment linkage in most fence units to ensure that the fence in front of the side head always stays parallel to the fence behind the head. Fence alignment is controversial, some men claiming that the front fence should 'lead-in' to the head, and others that it should 'lead-out'. If the principle of the surface planing is followed, i.e. two flat tables, parallel, but not in line, the two fences will always be parallel. The work then has the best theoretical chance of being planed straight. However, as the work has sawn surfaces, and as it may be out of true, some marginal gain may be made if a slight lead-in is given. Remember also that surface planing depends on skilled manipulation whereas the moulder controls the work with nip rolls and side pressures.

The out-feed or back fence behind the fence head is fixed and extends clear to the end of the machine. Both inner ends of the fence have adjustable pieces to keep the gap for the block to a minimum, and these pieces are often faced with wear resistant Stellite, or Meanite, a white cast iron. The lips of the bed plates are also hard faced.

THE PRESSURE SYSTEM

Throughout the machine there are side pressures. Starting at the in-feed there is one that keeps the work close to the fence before the feed rolls grip. There are usually two for the bottom head, one before and one after, although on the average machine the second one also acts to keep the work to the fence as it meets the fence head. These pressures are most often spring loaded, with adjustment for spring compensation and overall position.

The next horizontal pressure is that for the outside head. This must serve two purposes. It must hold work into the fence, and it must act as a 'chip-breaker' for the outside head. The block on this head will bring the work to width and its action tends to pull the work away from the fence. Furthermore, cutting into the end of the oncoming work, it tends to rive the grain deeply. Block and cutter design (see Chapter 2) can avoid some, but not all, break out. The pressure is made in the form of a shoe (hard faced) that is spring loaded to ride the work. As the sawn edge of the work feeds through it wavers in line according to sawing accuracy. The shoe rides this undulation and bites the edge immediately in front of the point of cut. Its precise position is adjustable to the cutting circle of the block and it is pivoted from a point diagonally across the block from its pressure point. This ensures that pressure is always as close to the cut as possible and that as the shoe rides out it clears the rotating cutters.

Behind the outside head the work is parallel; no spring loading is needed, so a fixed fence is adjustable to set width.

Top pressure performs the same job as side pressure, i.e. holding the wood steady and without vibration and movement other than forward feeding.

As the work top face is sawn it will vary in line and possibly in thickness. Top pressure must ride to this and still give control. The first pressure spans the bottom head opening. Most first bottom-head pressures are in the form of

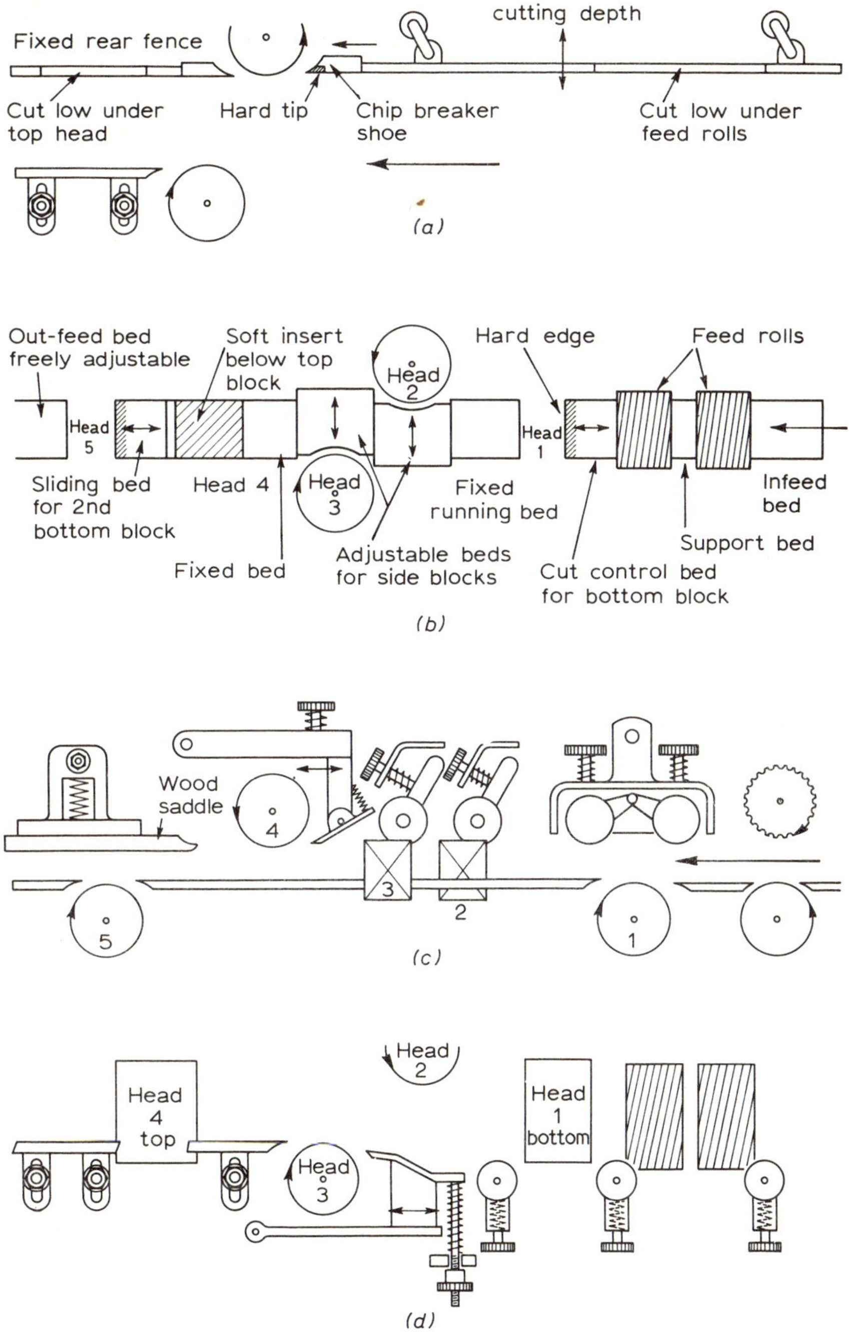

Figure 8.3 Fences, beds and pressures. (a) *and* (b) *standard arrangement of fences and bed platens for a standard four-head machine,* (c) *and* (d) *top and side pressures*

multiple rollers; each roller is sprung, but with the unit itself also sprung. Weight is incorporated by the use of heavy castings and the weight gives downward pressure with its tendency to 'jump' controlled by the springs.

There are top pressures opposite to each side head and in front of the top head. These top head pressures are massive in design and often incorporate a similar double-sprung system to that for the first bottom head. The pressure shoe of the top head pressure acts in the same way as that for the outside head and serves to hold work to the bed whilst it meets the cut and also to prevent chip break-out. After the top head there is usually a spring-loaded wood-faced pressure shoe.

The system outlined is for a four-head machine. Machines with more than four heads have similar pressures to suit each block.

Figure 8.3 shows fence, bed and pressure details.

DRIVES TO HEADS

Moulders have single-speed heads, two-speed heads or single high-speed heads. The smaller machines often have only one speed (not high speed) and are driven by 'V'-belt or short, endless, flat belt from individual standard motors (3000 rev/min) which are adjacent to each head. The pulley ratio determines the ultimate block speed. Power is mains 415 V supply, 3-phase, at 50 Hz.

The single-speed fast-running heads are driven by motors directly mounted to the block shaft, and these require frequency changers to give high speed. The two-speed machines have 'V'-belt or flat-belt drive and cone or stepped pulleys to permit belt interchange for speed selection. Some of these have high-frequency motors. Two-speed motors are used occasionally for this, but it means expensive multi-pole winding to give the correct ratio between the two speeds incorporated.

High speed is important to achieve acceptable surfaces and acceptable feed speeds. Surface finish is directly related to feed speed and cutter head revolutions.

The past standard measure of surface finish has been that of reference to 'cuts per inch', but with metrication the standard will be either the measure of the cuttermark pitch in millimetres, or the rather arbitrary 'cuts per 25 mm' or 'cuts per centimetre'.

Operating principles

The job must be assessed and cutter set-up planned. This entails consideration of the best running surface, e.g. face up, face down, and which edge to fence; the break-up of the cut between pairs of cutters (for clean cutting and good balance); the break-up of cuts between heads; and the condition of the sawn stock to be used (regularity and the amount left for planing).

When planning the running method, a flat running surface is aimed at although complete rounds may be run. Complete rounds run in pairs and split on the last block gives most chance of success for this work, but the set-up is more complex.

After the top head has cut, the top work surface must be held, so it may be necessary to plan for shaped saddles and pressures. Cuts spread between heads give easier and faster cutting but involve problems of registration if work-pieces are not straight and smoothly sawn. Large moulds cut with several pairs of cutters cut more easily than if cut with one pair of intricate cutters. Simple cutters are more easily made and more easily set, and if 'gaps' should appear, are more easily re-ground. Gaps in cutters are usually due to either

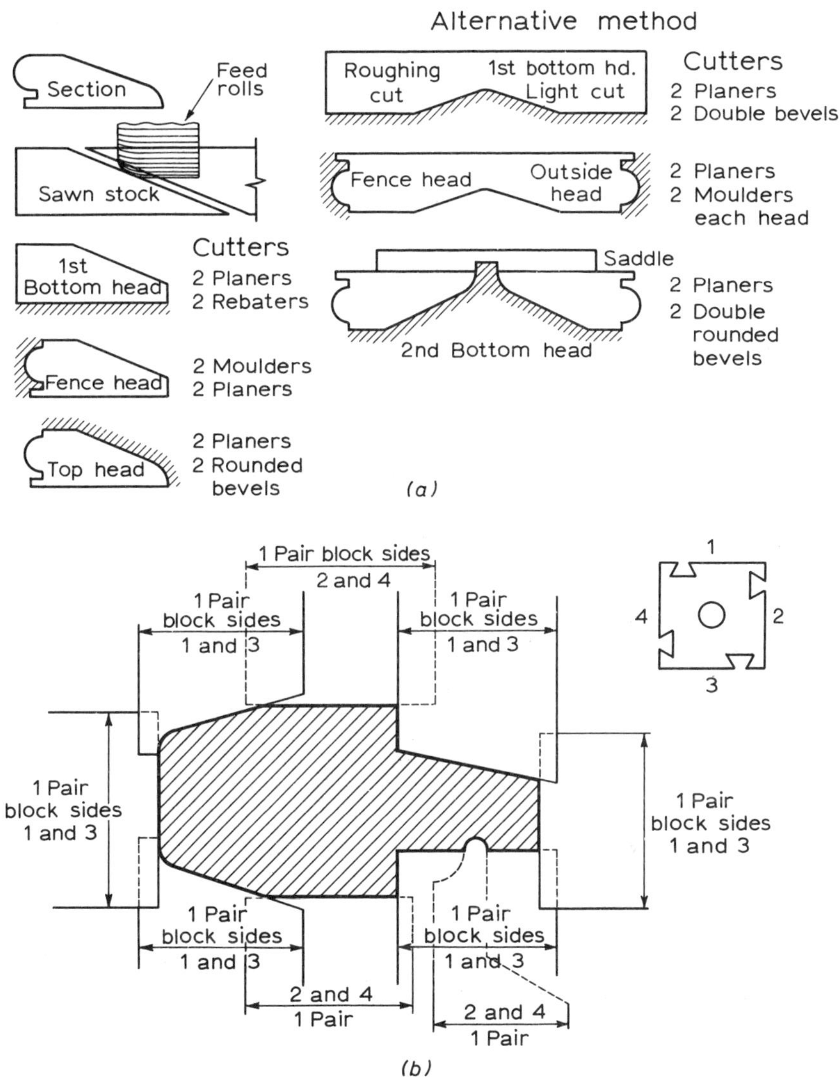

Figure 8.4 Set-up planning. (a) *Two methods of producing a common wood-trim section. Note that the outside head is not in use.* (b) *Common transome section and cutter set-up; perfect dynamic balance is achieved if cutters are in identical pairs*

grit in board ends or to overheating when being ground. However, a case can be made for cutting as much of a mould as possible on one block, with this one being the last bottom. With this method, since jobs are being run face down, no shaping is done by other blocks, only planing. The last block cuts downwards against the support of the machine bed. The job face is always clean, any saw marks left-on appearing on the top or back of the job. When changeover is necessary, only the moulding block will need changing. Additionally, once the work has past the last block it is almost out of the machine and requires little support and few shaped saddles. Another point is that when 'splitting' a job, i.e. running in double or multiple widths, the splitting cutters can easily cut up and into the top wood saddle without elaborate bed changing. Figure 8.4 shows variations in set-up planning.

Data

SIZES

Machine sizes are given by the maximum size work that can be planed. Standard sizes (in millimetres) are **75** × 100, and then **100** × 100, 150, 175, 200, 225 and 300; **125** × 175, 200, 225 and 300; **150** × 200 and 300.

FEED SPEEDS

Feed speeds are controlled by variable gearbox or multi-speed motors, and range from infinitely variable (from minimum to maximum) from 3 to 17 m/min on small machines and 7 to 33 m/min on medium machines to larger and faster matchers and floorers with speed ranges from 6 to 20 m/min up to 10 to 150 m/min. These are closely approximated as most machines are still rated in feet per minute.

POWER

Power required varies with machine size, and, as with all machine cutting, the load is directly proportional to the length across the cut. The longer the block the greater power it must have. However, machine designers have to design machines that can be sold, so the motors fitted are usually sufficient for average work where maximum section size, maximum depth of cut and maximum feed speed are not encountered. Larger and more powerful motors are fitted against customers' requirements. Top heads may well have larger motors than bottom heads. The fence side head may have a smaller motor than the outside head.

Motors range from 3·730 kW up to 29·84 kW.

HEAD SPEEDS

Cutter head speeds run from 4200 rev/min to 9000 rev/min. Most common two-speed machines have speeds of 4200 and 6000 rev/min.

9 Powered Hand Tools

Powered hand tools (often referred to simply as *p.h. tools*) are tools made to be hand held, and have a power unit operating a cutting tool. The cutting tools are related to the tools used on large machine tools, and do not often resemble normal hand applied tools. However, traditional hand tools are applied with skill, so too must powered hand tools. The use of such equipment demands newer, but just as exacting skills. Risk to the user is greater than with hand tools, but careful design by manufacturers and learned techniques by the user do much to minimise risk.

Power units

The power units built into powered hand tools are of several types. Electricity is the most often used power supply, but industrial tools are very frequently driven by the accumulated power of compressed air. Electrically driven tools have several types of motor, and even the type and nature of the electrical supply used, varies.

The drive unit does not materially affect the use of the tool; motors are designed to do the job for which the tools are designed and the user often has no choice but to use the tool for the job whatever the drive, subject to correct power supply being available.

Cutting jobs

Powered tools are available to do most cutting jobs in woodworking, and many jobs in general building and metal working. For hand woodworkers the tool types in use vary according to industry. Joiners may well use most saws and sanders; cabinet makers on assembly use most drills and screwdrivers; coach builders use circular saws, profile saws, together with drills, screwdrivers and nut runners. Tools are thus selected for a particular range of processes and users become skilled and proficient by familiarity.

SAWING

Sawing is by circular or narrow-strip reciprocating saw; there are no hand held bandsaws. There are a few models that use a long handsaw type of blade which reciprocates similarly to handsaws. Drive is by compressed air or electricity.

The circular saws in use are smaller editions of the larger diameter saws used on machine tools and are fully described in Chapter 1. There are saws with large teeth for rough rip and cross-cutting, saws with small, fine teeth for precision cutting, saws with hollow ground plates for planed finish sawing, and carbide-tipped teeth for abrasive sheet and laminate cutting. Power hand tool saws tend to cut a thicker kerf because of their need for greater stiffness. Circular

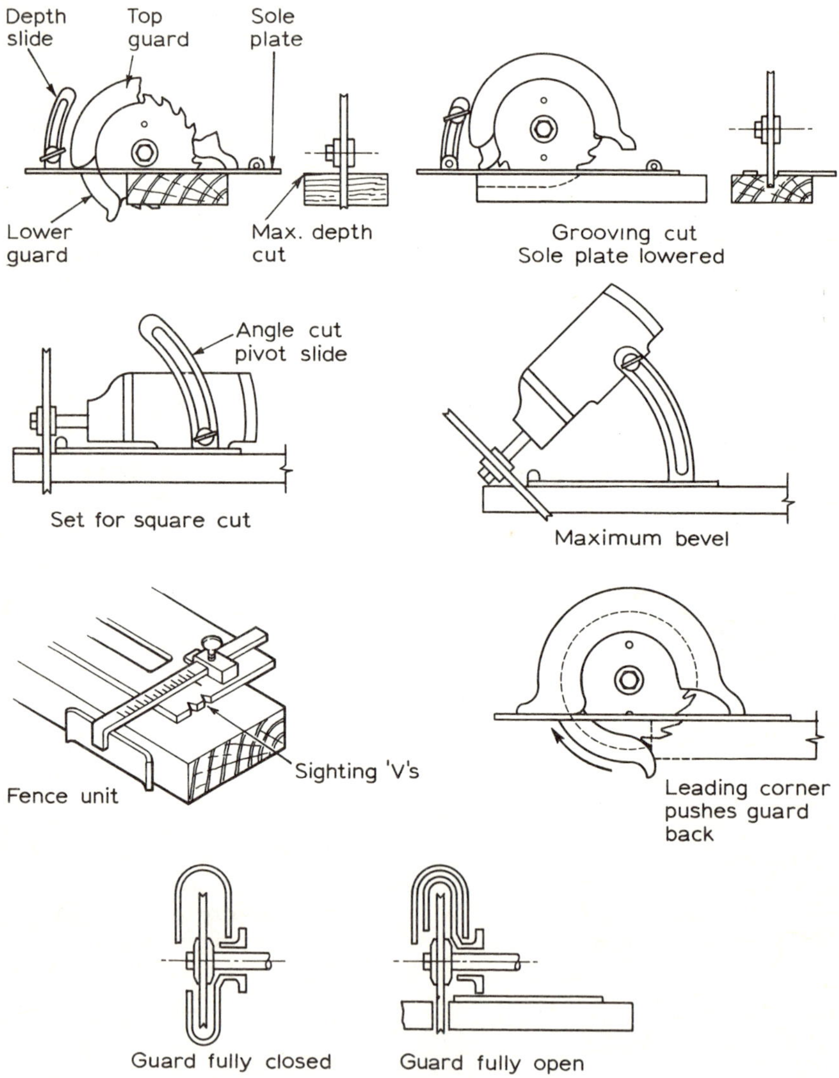

Figure 9.1 Saw unit

saws for power hand tools go from 100 mm to 250 mm diameter, with the most popular range being 150–200 mm. Some saws are coated with polytetrafluoroethylene (P.T.F.E.). These have a black coating with the proprietary name of Teflon. The purpose of the coating is to reduce friction in cutting and to avoid gum deposits.

The tool itself consists of the drive unit; saw arbor and collars, usually driven by geared drive but not always so; the sole plate, which on most tools will cant up to 45° for bevel sawing; the fence unit which acts as a guide that registers from the edge of a board for narrow strip cutting; a depth adjustment which enables saws to be used on their outer extremity (this gives the most efficient cut); and the saw guard. The guard is important, for most power hand tools continue to run for a short time after the trigger has been released.

The sole plate edges are parallel and can be used as guides when working across large sheets. Battens are fixed as fences for this. The front end of the sole plate has two 'V'-notches, one of which indicates the line of the saw for flat cutting, and the other which indicates the cutting line when the saw is tilted for bevel cutting.

The sole plate pivot is in line with the saw to maintain maximum possible cutting depth as the plate is tilted. For angle cutting the motor unit moves up and away from the sole plate. Main angles are usually marked. The sole plate is pivoted at its front end to enable cutting depth to be controlled. All cutting should be as near to the outer rim of the saw as possible. This gives the best surface finish, exposes less saw, and minimises the risk to the saw if the tool is twisted in cut. Variable depth cutting, together with the fence unit, enables rebates and grooves to be cut.

The saw guard is automatic. It consists of a curved channel section pivoted from the centre of the saw (around the saw bearing). The saw blade fits between the sides of the channel. When the saw is out of cut the guard is spring loaded down and around the blade portion below the sole plate. The guard front edge has a curved lip which meets the material to be cut and then rides the edge as sawing proceeds. When fully in cut the guard trails along the top surface of the workpiece until the saw emerges, and then swings down at once to cover the exposed part.

Figure 9.1 shows a typical power hand saw unit together with illustrations of work, and some working methods.

Profile Saws

These are compact tools for the purpose of cutting non-straight cuts. They may cut inside or outside shapes, and may penetrate by their own cut.

The motor unit drives a gear which converts rotary movement into a straight-line reciprocating one. The whole unit, saw and motor, rides on a sole plate. Models are available that have sole plates that tilt to one side to allow bevel cutting up to 45°. The front end of the sole plate is tip tilted up, and together with a slide facility (forward or back) this allows support close to the saw for variation in cutting requirements. Sheet laminate requires maximum support close to the saw, so the sole plate is moved as close to the saw as possible. If the sole plate is pushed back and a forward cranked saw used, cuts can be made into a flush corner. With the plate as far forward as possible and the front edges used as a pivot the saw can be raised clear of the work surface. When started and lowered the saw will cut into and through the work panel. An easier way for through cutting is to bore a hole for the saw.

To suit cutting demands, models are made with single speed, double speed, or infinitely variable speed. Trial and correction is the best aid to speed selection. Wood is cut most cleanly at the lower range of speeds, depending on thickness.

DRILLING UNITS

There is greater variation in these than is found in all the other power hand tools collectively. They start with minute air-driven models that are shaped to fit snugly into one hand, up to massive jobs that require a heavy two-hand effort. Drill sizes are rated by the maximum size hole that can be reasonably drilled in mild steel. This usually coincides with the chuck capacity, i.e. a drill rated as 13 mm will drill mild steel with drills up to that diameter, and as metal cutting drills usually have the same size shank as the hole diameter they drill, this is the chuck capacity. For hand held drills maximum metal cutting is 16 mm. The smallest chuck size is about 6 mm.

Powered hand drills vary most in their speed range. There are single speed, two speed, four speed and infinitely variable speeds. Single-speed drills can be high speed or low speed, and there are models that carry switching to convert them into very-low-revving impact drills where the drilling tool used receives up to 50 000 impact blows per minute, sufficient to shatter hard stones in concrete.

Drill Selection

Apart from the obvious, that the most robust tool possible should be used, the speed at which a power hand drill runs will decide its suitability for a particular purpose.

Cutting speed with drills is as important as it is with any tool. Metal heats up quickly with friction, and excessive speed will burn the drill point. Wood needs speed to sever the fibres cleanly. In terms of peripheral speed, wood is best cut when drilling at about 92 m/min, and metal (mild steel) at about 30 m/min. Drill speed S_D is given by the formula

$$S_D = \frac{\pi D R}{1000}$$

where D is drill diameter (mm) and R is the number of power hand tool revolutions per minute. The 1000 converts millimetres to metres.

To maintain these speeds an infinitely variable tool would be needed, because as the drill diameter is increased so is the peripheral speed. Again, even though power hand drill motors are designed to run at near constant speed under load, they do not necessarily do so; the larger the hole being cut and the harder the material, the greater will be the braking effect on the motor. A 13 mm capacity heavy-duty power hand drill will probably run at about 600 rev/min and be capable of drilling mild steel with 13 mm holes, and wood with holes up to 35 mm.

The feed speed of the drill is important. Metal must be sheared and this cannot be done under low drill pressure. The drill bit point becomes polished, the metal glazes and hard spots develop. Once penetration has been made, the drill should be kept under constant pressure until the hole is cut or the drill is withdrawn and re-entered. With wood this is not so important for the bits used have different forms.

Bits

For metal working, straight shank twist drills are used. For wood the drills are called *bits*. They vary enormously, and special ranges have been designed for power hand tools. All have the essentials for wood boring. (Tradition has it that hole cutting in wood is *boring* but industry generally relates *drilling* to small holes and *boring* to larger ones.) Wood boring requires a centre brad point to lead the bit on, a scribing point to sever fibres and a cutting edge to remove wood. Handworkers' Jennings bits are not suitable unless the centre lead screw is filed to a brad point, for the screw pulls the bit into the wood too quickly, or it clogs with dust and prevents in-feed. The helix angle of the bit flutes is too low for power drive work. For the general boring of wood the centre brad should extend sufficiently beyond the spur to centre accurately the bit. Where curved surfaces have to be bored (chair legs for dowelling) the brad must be long enough to centre firmly before scribing shoulders contact the surface. The same principle applies to holes bored at angles, the brad must anchor first.

SANDERS

Sanding is the name given to the job of abrading a surface with a continuous field of abrasive points. These are the sharp edges of abrasive grit that is bonded to a backing. The characteristics of coated abrasives are covered in Chapter 10.

The methods by which the abrasive points are passed across the wood surface vary. There are three main methods. The first has the abrasive-coated cloth or paper made up into belts and these are run around the two pulleys of the power hand belt sander. The second most common is the orbital sander where the coated abrasive is held in the form of a flat pad and given a circular motion of about 5 mm diameter; these are mainly used for finish sanding as they do not remove much wood. The last of the power hand sanders is the disc. Here a cloth-backed disc coated with abrasive and backed by a flexible-rubber disc pad, is rotated by the tool motor.

Belt Sanders

Belt sanders run in size (by belt width) from 65 mm up to 100 mm, but heavier models for floor and bench sanding go up to 150 mm. They can be paper or linen backed, the latter being most serviceable and most preferred, for they tend to wear out or clog and be changed before they break, i.e. before they are fully utilised.

Most sanders have at least one pulley of rubber and this is the driver which is driven by a chain from the motor through a gearbox. The pulleys are flat and are set above the line of the platten, which lies between them. This platten is the working surface and consists of a resilient pad of foam rubber backed by a rigid plate. In front of the rubber there is a metal shim plate and in front of this again is the back of the working portion of the belt.

Belt tension is by tension spring with a snap locking device that maintains tension as the belt stretches. To keep the belt on the wheels and to allow for it to be tracked to the centre, there is a tracking control. The sander is held bottom face up with the belt running, and the adjusting knob is rotated until the belt runs to the centre. The adjusting control increases or decreases the degree to which the pulley centre lines are out of parallel. Very few belts will run on pulleys that are completely parallel, for belt edges stretch, and they are difficult to make with both edges exactly the same length. The out-of-parallel pulleys accommodate this. Flexible belts tend to run to the lowest part of the pulley, so to move the belt to the inner side of the sander the pulley axis lines are closed at this end. To pull the belt towards the outer side of the machine the pulley axis lines are closed at the outer end.

Sanders should be started before contact is made with the work, for slow running abrasive particles scratch deeply. Likewise, the tools should be lifted away before switching off.

The design of the power hand belt sander allows the belt to be run close to rebated or upstanding edges. Incorporated in power hand belt sanders are extraction fans. These pick up dust as the belt leaves the work surface and direct it through a tube into fine weave linen bags, where it is trapped, and the cleaned air allowed to escape through the weave.

Belts are joined by scarfing the abrasive coating and lapping the backing. This leaves a slight 'plus' thickness that leads to slight bumps as the belt runs. Any lap will have one edge leading to the belt face. This edge should 'trail' to the work to prevent pick up. To assist with this, belts have direction arrows printed on the inner face of the backing.

Orbital Sanders

These orbit to cut; they are used mainly for finishing, whereas belts are used for rough cutting and finish preparation. The abrasive used on orbitals is mainly paper backed and is either bought in graded packs or cut from standard hand-work papers.

Orbital sanders have resilient faces to their orbiting platens to cushion the abrasive paper. Their movement is circular, but some later machines have controls that allow them to be switched to straight line (forward and back) oscillation for fine finishing.

Air-powered pad sanders mostly have two pads on the larger sizes, and one smaller pad on single hand models. Double pads oscillate from front to back and do not orbit.

Disc Sanders

These are used for shape forming and for first cleaning prior to hand or machine finishing by other methods. The disc can be presented to the work at almost any angle and is suitable for chair-arm sanding and for the edges of preformed ply shapes. They are not suitable for finishing because the paper disc edges tend to cut in and to leave circular marks. The paper discs clog very quickly, as they run at slow speed near the centre.

Figure 9.2 shows general illustrations of a range of drills, profile saws and sanders.

PLANERS

Planers are small-scale replicas of machine-tool surface planers, with the difference that the power hand planers ride the top of the stationary workpiece.

The principle is the same: two flat and parallel tables (sole plates) are separated by a circular cutter block. The lead table controls the cutting depth and is adjustable. The back table is set level with the line of cut of the cutters. With skilled application straight edges can be formed. High spots are reduced first and then continuous through cuts produce the required straight line. With surfaces wider than the length of the cutter it is difficult not to leave ridges between cuts, but with light cuts and cutters that have their corners rubbed down (or rounded), acceptable surfaces are left for sanding later.

Fence units are sometimes fitted; these are most useful where the cutter block runs right up to the edges of the sole plates. This allows rebates to be made. With a fence that pivots, bevel cuts are possible.

To achieve good results, cutters must be kept sharp, and in use the planer must have firm pressure applied to the front sole plate until the cutter starts to cut. As feed progresses, pressure must be transferred to the rear sole plate. This technique is derived from hand plane manipulation.

ROUTERS

Routers are very popular with small joinery workshops because their work is almost without limit, providing there is a full supporting range of cutters and equipment, and that there are men practised in their use. The work range includes rebating, grooving, chamfering, bevelling, housing, moulding, pattern shaping and rough dressing of compound shape work and laminated shapes. They are used to cut the compound bevels on the edges of planking for carvel-built boats, and for port hole windows in composite sheet material. Tools are available that enable plastic laminate trimming and bevelling.

The router consists of a motor unit, the armature shaft of which forms the spindle nose for the attachment of tools and a clamp frame that has hand holds and a circular base plate. In addition there are clamps for securing a fence unit. Adjustment is provided for controlled movement to regulate cutting

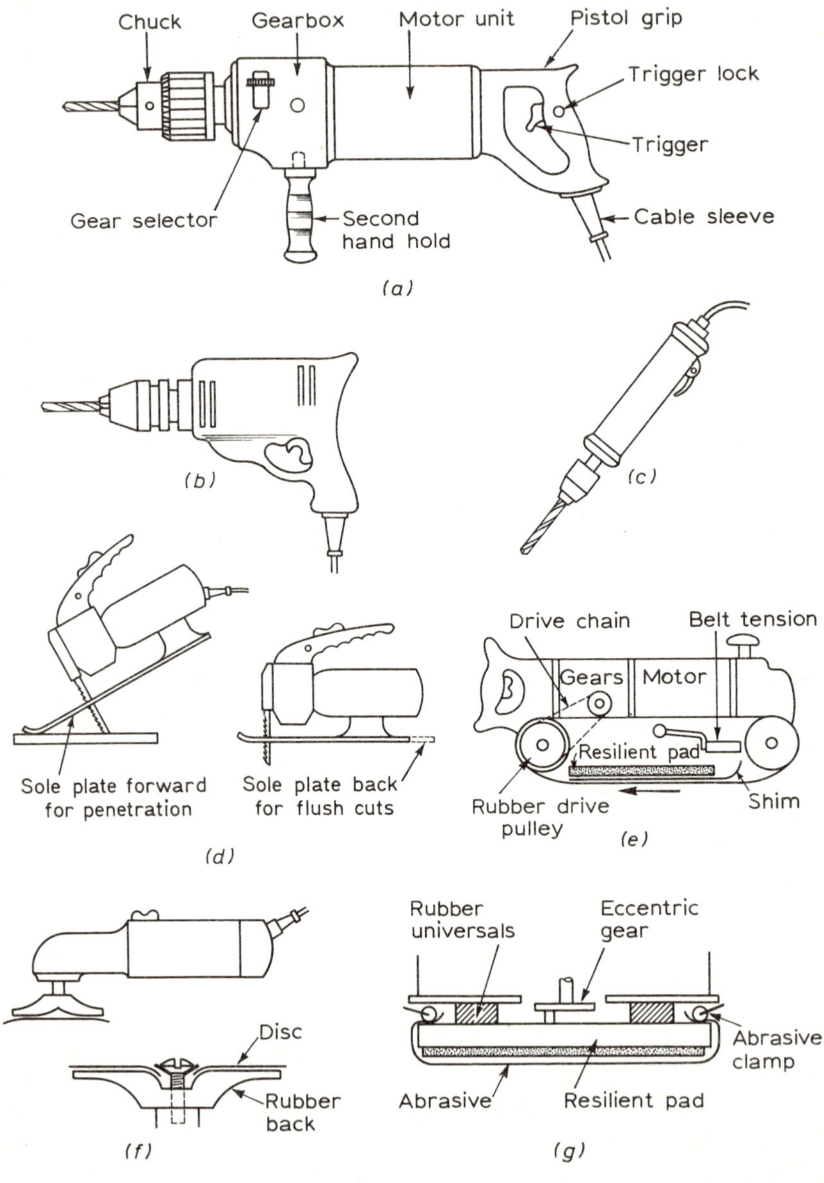

Figure 9.2 General range of powered hand tools. (a) *Heavy duty drill (up to 16 mm)*, (b) *single hand drill*, (c) *Vane motor air drill*, (d) *profile saw*, (e) *belt sander*, (f) *disc sander grinder and* (g) *orbital sander*

depth. A quick-release clamp frees the router body to slide within the clamp frame and this enables quick setting for the approximate depth of cut. Precise cutting is achieved if the body clamp is made tight and the fine movement adjuster used. There is also a means of locking the spindle for cutter changing.

Cutters are mostly held in collets which are split and tapered. According to the size and design of the tool the collets can be locked tight on the cutter by an enclosing nut screwed onto the threaded spindle nose (these push the collet up into the hollow spindle), or, on some larger power hand routers, the collet is pulled up by a top screw which passes down through the hollow centre spindle. Some power hand routers have all the cutters with screwed shanks which screw into the spindle nose.

Cutters are similar to those for machine tool routers, although the eccentric chuck principle is never used. The main rebating and grooving cutters are double-edged panel cutters. Smaller sizes are single-edged spoon bits. All have back relief, or clearance.

For moulding, small solid-profile cutters are used and these are two edged with a shear angle of cut. Edge grooving is done by small solid groovers rather like circular saws, but having only 2 or 3 teeth, and with solid centres and shank. Arbor adaptors can be used and on these either small slitting (circular) saws are fitted or solid profile cutters having a centre hole. For laminate edge trimming there is usually a three-wing carbide-tipped solid-profile cutter, square or bevelled. Below this cutter is a small ball-bearing roller on the arbor. This roller rolls along the panel edge and the cutter trims flush as it moves along. Peculiar to the power hand router is the solid profile cutter that has a centre pilot pin which extends beyond the lowest cutting point of the cutter. This acts as a runner, allowing the tool (and cutter) to be traversed along the edge of the material being cut, and without any other form of guide. However, feed must be continuous and with no undue pressure. The high speeds of router cutters quickly generate frictional heat and burn the work and cutter if pilot cutters are traversed too slowly and with too much pressure.

Figure 9.3 shows details of the power hand router.

Working Method

Routers must have controlled guidance. They are not intended for free hand cutting. Very small cuts can be hand guided for rough carving, but otherwise the practice is highly dangerous because of the risk of snatch. Router guides must be in contact with the work surfaces before the cutters start to cut.

Guides can be fence units extending from the router base. The edge of the base itself may sometimes be used. For arc cutting, the base can have a plywood arm extended which is pivoted to give the correct radius of swing. Port hole windows can be cut in this way. The plywood arm is pinned to the hole centre and the router rotated and lowered until approximately half the depth is cut. Both sides are treated in this way, leaving a thin web to support the cut away portion. Final finishing is by hand.

Cutters having pilot bushes can be used to run small edge-mouldings.

The method of most general application is that of using templet guides. These are circular guides that fit to the router sole plate. The cutter runs down through the hollow centre of the guide and extends into the work.

For straight housing, battens are fixed across the work. The router sits on top of these and the templet guide runs in contact with the batten edges. Batten spacing is decided by cutter diameter, templet guide diameter and the width of the housing required. To cut a housing 25 mm wide using an 18 mm cutter with a 50 mm templet guide, the batten spacing would be templet guide

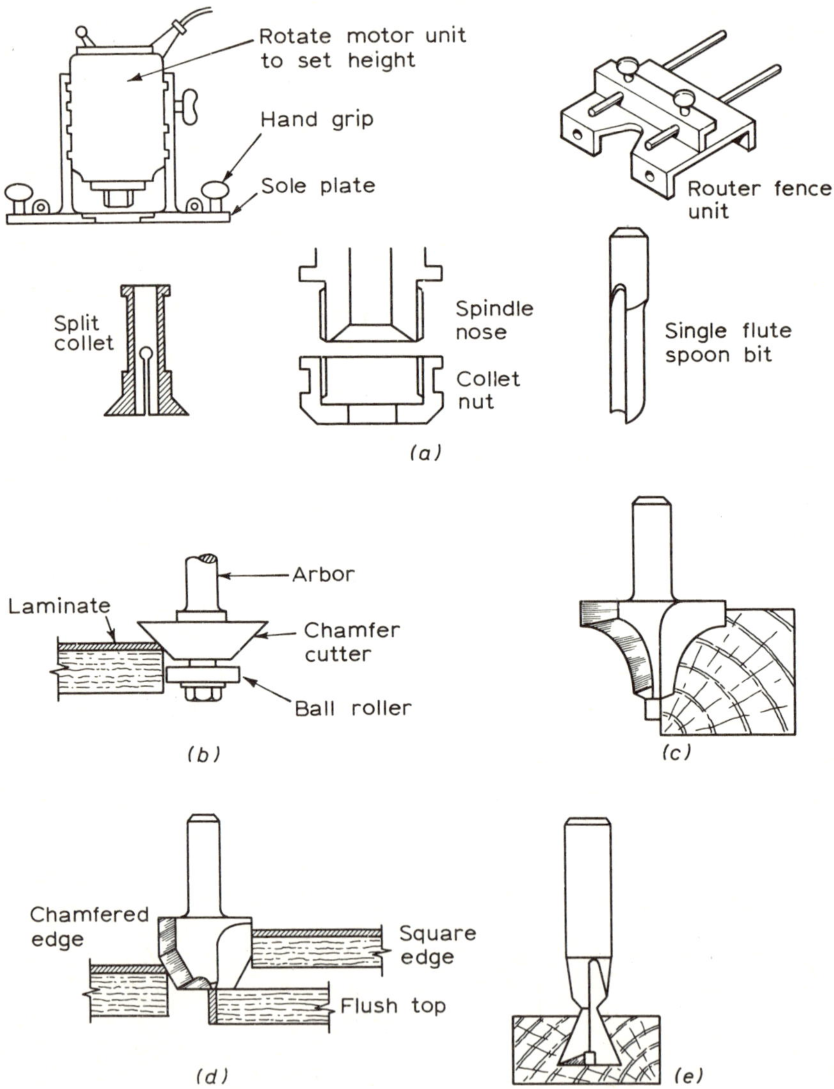

Figure 9.3 Power hand router. (a) *Main features,* (b) *laminate trimming tool,* (c) *solid profile cutter (ovolo) with guide pilot,* (d) *combination form of laminate trimming cutter (the router fence unit must be used with this tool)* and (e) *dove-tail cutter*

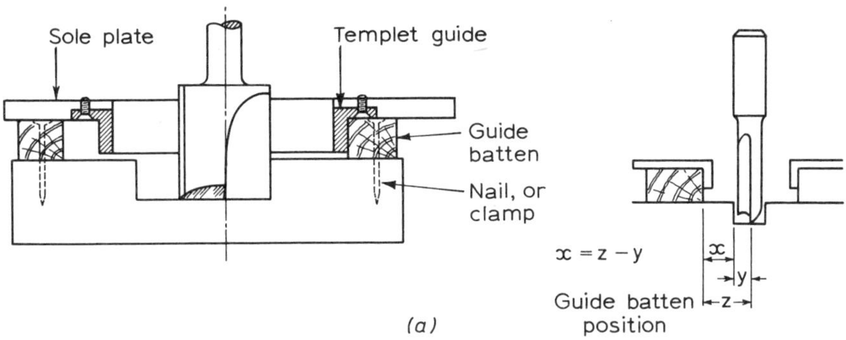

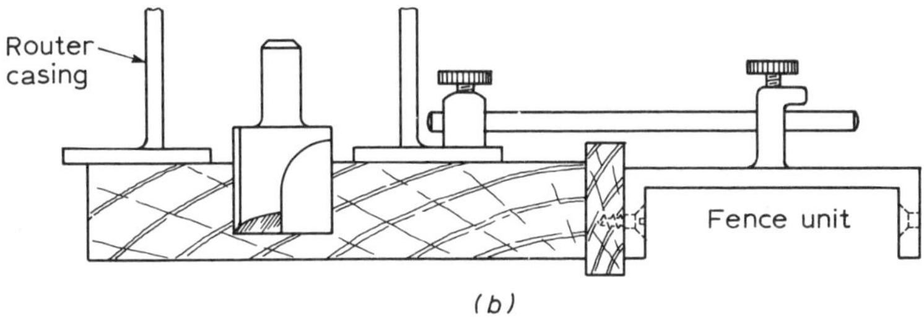

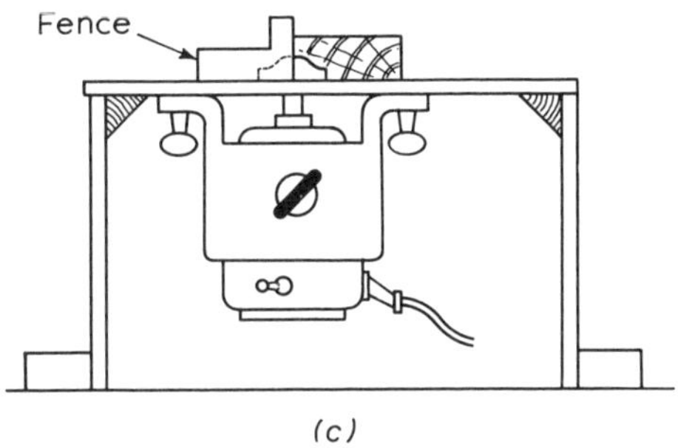

Figure 9.4 Router applications. (a) *Application of templet guide to housing, and method of finding a guideline,* (b) *ploughing a wide groove using the fence unit as a guide and* (c) *router up-ended and attached to a purpose-built table for use as a small spindle moulder*

diameter + required cut − cutter diameter, i.e. 50 mm (templet guide) + 25 mm (housing) − 18 mm (cutter) = 57 mm, which would be the correct spacing between guide battens. The cutter diameter must always be at least half of the required housing width.

If a templet guide is used to cut a profile shape where a templet is attached to the top of the material to be shaped, then calculation is concerned with the amount by which the finished edge will overlap the templet.

$$\text{Overlap} = \frac{\text{Templet guide dia.} - \text{Cutter dia.}}{2}$$

i.e. half the templet guide diameter—half the cutter diameter. A 300 mm port cut through block board would need a templet cut over size by twice the overlap distance.

With a 12 mm cutter in a 50 mm templet guide, the overlap of cut edge over templet edge would be

$$\frac{50-12}{2} = 19 \text{ mm.}$$

This would also be the amount by which the radius of a curved templet for an internal curved cut would be in excess of the radius of the curve to be cut.

Cutting

Cutters must run towards the work. This means that the rotating cutter will pull the power hand router snug into the templet and the cutting track is then smoothly controlled. If cutter climb cuts then the tool tends to snatch and run away from the guide. The general path of the router must be clockwise for inside or perforated shapes, and anticlockwise for outside shapes.

Cutters must be kept sharp by grinding, using small grinding wheels that are round nosed for fluted cutters and that are of the flaring-cup type for double-edged panel and profile cutters. All cutters are ground on their internal face, never on the outside. Small attachments are available for grinding. The grinding wheels are attached to the router nose, and the router is then held bottom-up in its clamp. A fence unit is used to guide a cutter-holding grinding attachment across the base of the sole plate and past the grinding wheel. Fine adjustment is provided for rotation of the cutter to ensure maintenance of the cutting angle.

Figure 9.4 shows applications of the power hand router.

SCREW DRIVERS, EDGE TRIMMERS AND HAND JOINTERS

Screw drivers are slow runners. They have a standard drill body and motor unit, but have a reduction gear to give the correct speed for screw running. In place of the drill chuck is a sleeve guide that contains the driving bit. This must be changed for each screw number, although smaller screws than the sleeve size can be run. The main part of the tool is the clutch. This consists of two parts. One is a simple spring loading that engages the main drive when the bit is pushed back by its pressure on the screw. The bit does not rotate until sufficient pressure is applied to operate this part of the clutch. The screw is then run in until the required depth is reached. At this point resistance to turning builds up sufficient torque to disengage the second part of the clutch.

Adjustment is provided for the second part of the clutch so that screws may be left proud, flush or deeply countersunk. The adjustment has a screwed ring that compresses the clutch spring more or less, as required. If this range of adjustment is insufficient some screw drivers have interchangeable compression springs, colour coded. The heavier the spring used, the deeper it will be possible to drive the screw, or the larger will be the screw that can be driven. If the screw is driven too deeply, the driving bit slips, spoiling the screw head and defacing the wood around it.

Some drivers have two-speed boxes and others have two-speed ranges with infinite selection by trigger pressure. Some have a reverse gear to allow for out-running of screws.

Wood screws are run in slowly; self tapping (parallel thread) screws are run in fast. Speed selection is by judgement related to the size of the screws and the material into which they are being run.

EDGE TRIMMERS

These are one-purpose tools to trim plastic laminate sheet back flush with the base board to which it has been applied. The edge finish can be square or chamfered according to the cutter used or the position of the cutter relative to the work. The cutters are double-edged tungsten-carbide tipped. Tungsten carbide is a very hard wearing metal tough enough to stand work with resin glue and abrasive laminates.

Some cutters have two square edges, others two bevelled edges. Some have one edge bevelled and one square. The square edge is longer than the bevelled one and cuts to a smaller diameter. The bevelled edge, set above the end of the straight edge, acts to a much greater diameter. For square trimming, the cutter is set with the square edge into the cut; for bevelled finishing the cutter is lowered until the selected amount of bevelled edge is engaged.

The tool has a standard motor, and a standard drill or router chuck. At 90° to the cutter axis there is a flat work guide. This is held in contact with the panel face and guides the tool smoothly along. Adjustment of the fence towards, or away from, the end of the cutter controls working depth. Opposing the lower end of the cutter is a small ball-bearing roller. This is positioned face to face with the cutter end, both axis lines running parallel but not in line. The roller edge lines-up with the cutter edge and runs along the edge or face of the panel being trimmed and guides the tool. A fine screw adjuster controls the degree to which the panel is trimmed flush.

THE HAND JOINTER

This is another one-purpose tool, but one with an enormous use potential. Particle board, plywood and block board often have to be joined together, flush, offset, edge to edge, end to end, mitred at right angles, and at square corners.

Because tongues and grooves are not satisfactory in multi-layer particle board, short gashes and tongues are often used for location in the furniture cabinet industry. These are cut on very expensive and very automatic double-ended tenoning machines. Small scale producers and other semi-hand producers cannot easily utilise this method.

The hand jointer* enables such sheet material to be joined easily. The tool has a motor unit and a flat base plate on which it can be easily slid along. The cutting tool is a 100 mm diameter carbide-tipped saw, cutting a groove 3·5 mm wide. This is driven through a right-angled gear drive and lies parallel to the flat base. At the end of the base there is a small fence rising at 90°. The saw is behind this fence, which is spring loaded. If the tool is pushed along on its base until the fence contacts the edge of a particle board panel, the fence is pushed

*Manufactured as the Lamello–Minilo and sold by J.K.O. Cutters Ltd.

back. As it goes back the saw emerges and cuts a groove or notch. The fence has a pre-set stop that limits the depth of cut. If several of these crescent-shaped cuts are made along an edge, and another series along the face of another panel, the basis of a right-angled joint is prepared. Into each gash a compressed wood key is glued. The two boards are assembled together making a right-angled joint.

The method is quick and accurate, and only requires simple jigs to locate tool and work. The keys must be purchased in bulk, and are compressed so that they can swell tight on absorption of glue.

Figure 9.5 shows the principles of the design of the screw driver, edge trimmer and the hand jointer.

Electrical details

Motors for powered hand tools may be mains (single phase or three phase) or transformer low voltage, all alternating current. Direct current is not commonly used, but most motors are a.c./d.c. although the trend is towards winding for a.c. only.

The main range of tools are single-phase a.c./d.c., and are described as 'series-wound universal motors'. The 'universal' part of the description means that they may be used on varying frequencies from 25 Hz to 60 Hz, and also on d.c. Their voltage acceptability is limited to $\pm 5\%$ of the indicated voltage. They are essentially fast motors, running best at over 3000 rev/min. Modification of their internal winding will bring the speed down to 2000 rev/min.

With universal motors, the rated speed indicated will only be achieved if the load is rated, i.e. a 6 mm power hand drill will run at rated speed when used to drill a 6 mm hole in mild steel. If heavier work is given, the motor slows; if lighter, the motor accelerates. This means that universal motors of this type are best used intermittently, e.g. for non-continuous work, and where constant speed is not essential. Brush wear is high, and when in continuous use temperatures rise fast. Electrically, the motor is known as a 'series'-wound motor.

In Chapter 7 it was said that the turning effort (or torque) of a squirrel-cage induction motor on three phase was caused by a rotating magnetic field set up in the field windings. The magnetic field rotated because each of the three phases reached peak voltage 120° after the preceding one. This electromagnetic field interacted with the induced field in the rotor to produce torque.

Single-phase series-wound motors have their rotating fields set up in a different manner. Unlike the squirrel cage, the moving part of a series-wound motor, i.e. the armature, is wound with coils. The non-moving part, i.e. the field, is also wound. The coils of the armature are connected end-on (in series) with the field coils through a device known as a 'commutator and brush gear'.

Field coils are wound round packs of iron laminates to produce maximum magnetism and minimum electrical eddy currents and heating. Two poles are usual. One field is wound to produce a 'north' pole. The inner end of this coil goes to a carbon brush that runs in contact with a copper commutator. One commutator segment is connected to the end of a coil wound on the armature to produce a 'south' pole. From the inner end of the south pole coil a connection

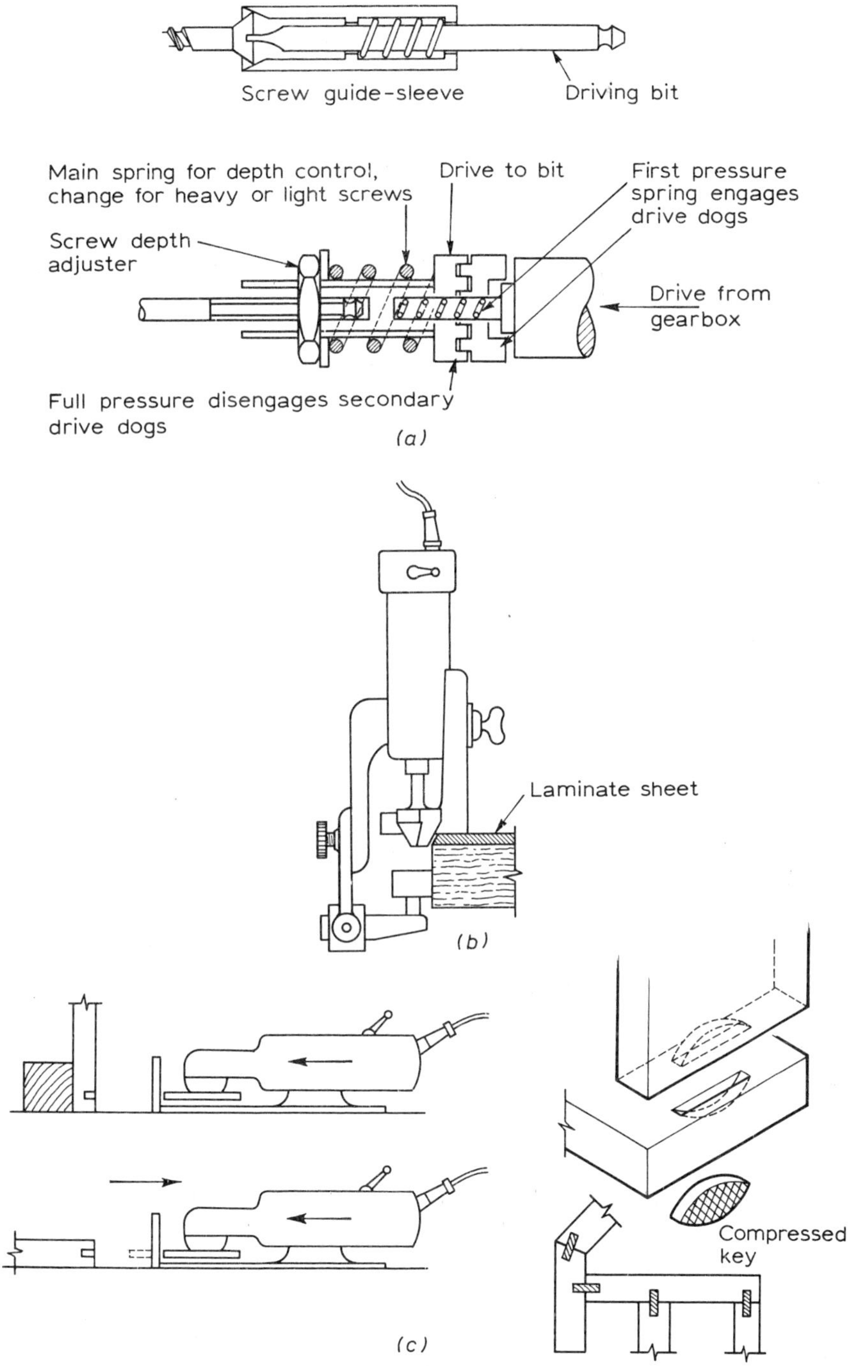

Figure 9.5 (a) *Screw driver clutch,* (b) *edge trimmer and* (c) *hand jointer*

is made to another coil to produce a 'north' pole still in the armature. From here the coil is connected to the other commutator segment which has a second carbon brush. This second brush connects with the second field coil wound to produce a 'south' pole (in practice, armatures are wound with more than two coils). When the power is connected, the field north repels the armature north, which is opposite it, and the south field repels the south armature pole. This creates torque and the armature rotates. The armature north and south are then attracted by their opposing field poles (north to south), opposite poles being attracted. When the two opposing poles are adjacent, rotation would stop (this is the basis of dynamic braking—see Chapter 4) but at about this moment the commutator segments, through the brushes, have changed the polarities of the armature north and south poles. (An iron rod, wound with a coil through which a current is passed, has a north and south end, according to the direction of flow of the current.)

Motors that are a.c. only are identical, except that they are without the extra coil that is put in for d.c. running. A.C.-only motors are smaller and develop more power, size for size, than a.c./d.c. motors.

LOW VOLTAGE WORKING

Low voltage working is used in the interests of safety. On building sites, tools are subject to heavy wear and tear. Production lines in factories have the danger of several supplies to each work space, e.g. lighting, power, single-phase, three-phase, etc. and danger is present. Low voltage is safer because it is the element of voltage that produces 'shock' in human beings. Reduced voltage therefore minimises risk should breakdown occur. The voltage selected for power hand tools is 110 V. This can shock, but is usually non-lethal. Coupled with a particular form of wiring it can be restricted to half (55 V) if breakdown occurs.

The system has a transformer, which reduces alternating voltage by induction. The primary (or input) coil takes mains voltage. Any coil carrying a current will induce (by electromagnetic induction) a current in any adjacent coil. This secondary current will be inversely proportional to the primary, according to the relative number of turns of wire in the two coils. A 10:1 reduction will have one coil in the secondary for every 10 in the primary. The voltage drops in proportion to the number of coils in the secondary winding.

The transformer takes mains a.c. at 240 V and reduces it to 110 V, which is suited to low-voltage-powered tool motors. So far, any contact with the secondary circuit (the tool) will give a maximum shock of 110 V. The centre turn of the secondary in the transformer is now connected to earth. If a break or short occurs anywhere in this circuit there is now a maximum potential to earth of 55 V. Low voltage tools must be clearly marked, and special plugs and sockets must be used. It must not be possible to mix supplies through careless plugging.

Pneumatically powered tools

Pneumatic, or air driven tools are not as common as electrically driven ones.

For wood they are mainly drills, screw drivers and sanders. They are more expensive, being better engineered because of the necessity of preventing air leaks, but power for power they are smaller than electric tools. They are more simple in design and require less maintenance.

Air motors vary, but the vane is the usual type for power hand tools. This has a central rotor that runs in a cylindrical bore. The cylinder is the stator and has two ports in its walls; one for the infeed of compressed air, and the other for the exhaust of used air. The rotor lies eccentric within the bore, and although its diameter is much smaller than the bore diameter, the wall of the stator almost touches the wall of the rotor in one place. The air-inlet port is close to the point where the stator and rotor are close together. Set in slots in the rotor are a number of vanes which are of plastic or soft metal to avoid wear on the walls of the stator. These vanes can be pushed down flush with the rotor surface or they can slide out until they contact the stator wall. When the rotor rotates, the vanes are flung out as far as they can go, and in one revolution they will move from minimum projection to maximum, and back to minimum.

When compressed air enters the inlet port, some of it is fed to the rotor end behind the vanes where it pushes them out to create an air seal against the stator. Being eccentric, the rotor and vanes form a number of enclosed cavities which are roughly triangular in section. Compressed air expands to fill any available space; it then thrusts with equal pressure on all enclosing walls. The incoming compressed air is fed to the most restricted part of the cavity between rotor, vane and stator. It enters and expands, pushing against the vane at the largest end of the cavity with greater force than it pushes against the vane at the smallest end. This is due to their differential areas. The rotor moves and the air continues to expand as the cavity enlarges. When the point is reached where the rotor starts to get close to the stator wall again, the outlet, or exhaust port, allows the air to escape.

Thrust on the motor is proportional to the air pressure which is usually in the range 0·53–0·66 MN/m^2*. The speed of rotation increases as the air pressure increases and torque increases at a similar rate; there is almost no starting torque.

Air tools are perfectly safe. If stalled by excessive load no damage occurs. Because of their rapid acceleration they should be in light contact with the work when the air valve is opened.

Powered hand tool maintenance

Most tools have 'packed for life' bearings. Some have oil nipple lubrication. The first require no servicing maintenance until the tool is returned to the maker for complete overhaul after long use. The second range require an occasional drop of recommended oil.

Gearboxes are prepacked with light grease which should be cleaned out and renewed if any user overhaul is undertaken. The correct grade of grease is usually listed on the tool instruction leaflet.

*This is still frequently referred to as 80–100 lbf/in^2, but an acceptable S.I. alternative is 5–7 bar. However, BS 3763: 1970 states that the latest recommended S.I. unit of pressure is the pascal (Pa). 1 Pa = 1 N/m^2, and 1 bar = 10^5 N/m^2.

Air tools should be automatically lubricated through the air system. Compressed air from the mains should be filtered for dust and water, and lubricated, before it enters a working tool. This is done by a wall unit called a 'lubrication and filter set' or an 'air set'. Filtering is by centrifugal swirling, and lubrication by droplets of oil which are atomised and carried on by air flow. All parts of the air powered tool are reached by this type of oil mist lubrication. Apart from lubrication, air motors need vane changes and sometimes air seals. Screw fittings and adaptors should be checked for leakage and air hose inspected for cuts.

Another aspect of maintenance for electric tools is cable inspection. Cables must be free from cuts and damage and must be firmly fixed by cable grips inside the tool and plug. Rubber sleeve protectors at the tool and plug must be seen to be in good condition. Any loose screws must be tightened and moving parts of attachments lubricated.

Brushes, which are carbon blocks, need replacement when worn. Those that are badly worn, cracked or poor fitting tend to spark. This burns the commutator away. New brushes must be the correct ones for the job and must fit easily into the tool. Spring tails of brushes should not be stretched and spring clips must not be bent. Commutators should be rubbed clean with a dry rag; the use of solvents may ruin tool insulation.

Brushes must run lightly in contact and move freely in their holders. A fine line of minute sparks as the armature rotates is acceptable. Compressed air should be used to blow dust from the field and armature coils.

Safety

The use of any power hand tool is potentially dangerous. Care, skill and observation of safety procedures will do much to prevent accidents.

The Factories Acts (Electrical Regulations) of October, 1968, cover the design and use of powered tools. In addition, the Employer's Liability Act, 1969 (Defective Equipment) imposes a duty on the employer to maintain equipment in good order.

Basic safety is ensured by adequate servicing that gives attention to cables, plugs and sockets.

Switchgear should be double pole, breaking both lines, and all connections, both internal and external, should be firm. The overturning of screws may shear fine copper strands of the cable and thus lead to shorting and defective working in the future.

All circuits should be fused to the capacity indicated in watts on the tool. Fuses are rated in amperes. The current I (in amperes) is related to the power W (in watts) and the voltage V (in volts) by the expression

$$I = \frac{W}{V}$$

A tool rated at 800 W on 240 V should have a 3·3 A fuse, but for practical working, a 5 A fuse would be quite safe.

All metallic parts must be fully earthed except for 'double insulated' types.

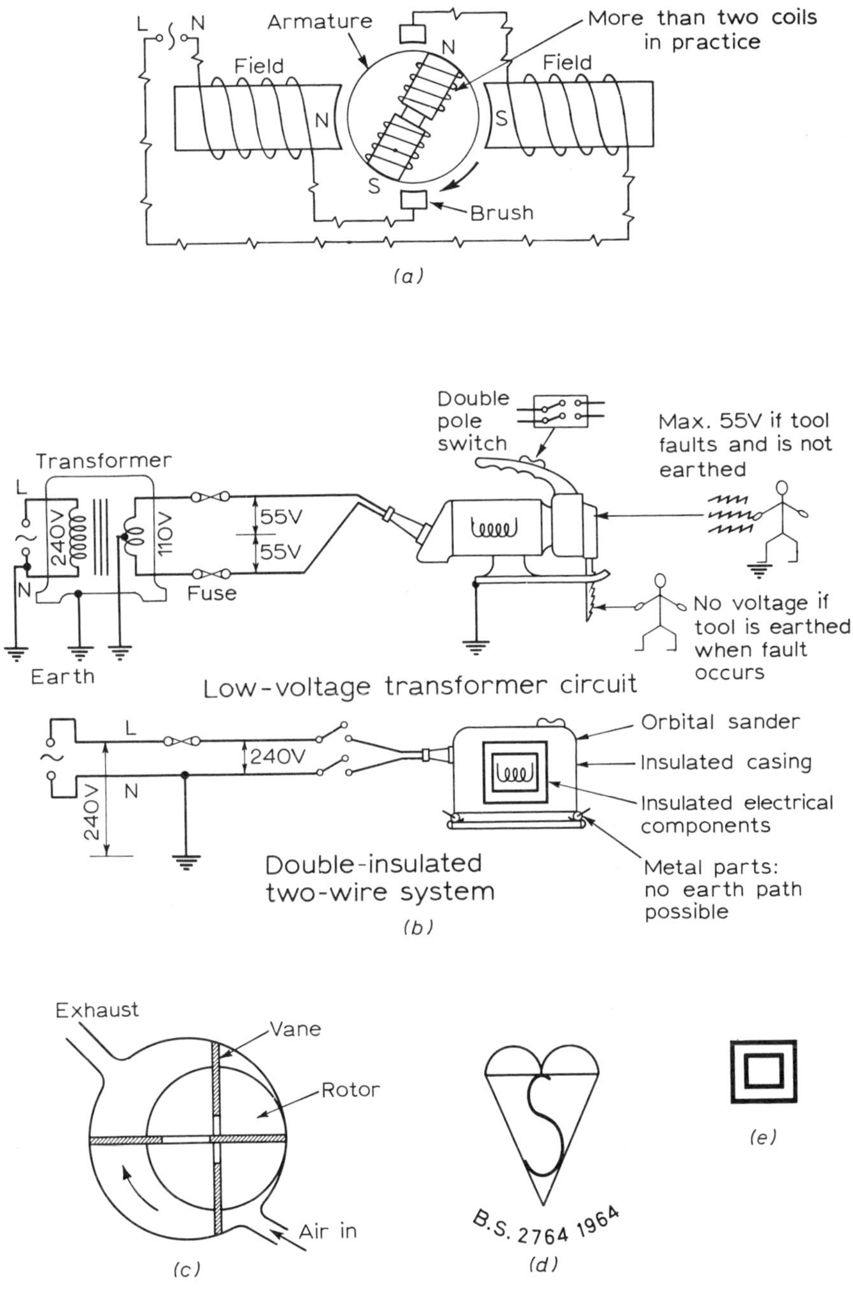

Figure 9.6 (a) *Series-wound motor,* (b) *wiring circuits,* (c) *Vane air motor,* (d) *British Standards Institution Kitemark and* (e) *double-insulated mark*

Earthing ensures that if mains voltage is connected to an earthed part, current will go directly to earth and the fuse will blow.

Briefly, all double-insulated tools have two-wire connection and no earth. All electrical parts are fully encapsulated in insulating plastic within the enclosing casing, which itself is made of a tough, shockproof and non-conductive plastic. All external metal parts are insulated from the live motor. The cable remains the weak link and so low voltage working is advised by the Department of the Environment.

General safety rules apply to all power hand tools, e.g. no adjustments without disconnection, full use of guards on saws and planers, use of goggles if particles of waste are flying. Tools must not be pulled by their cables, which must be of the correct current-carrying capacity. Each electrical type of machine must have its own distinctive system of plugs and sockets.

Traditional wiring has the live lead coloured red, the neutral lead black, and the earth lead green. Double-insulated tools will have only two wires, i.e. red and black. The standard with which all new tools comply has a different colour code. Here the line (or live) wire is brown, the neutral is pale blue, and the earth, where fitted, is green/yellow (striped).

In use, all tools must be kept sharp and not forced. They must be supported until they cease to revolve.

Speed control

Many tools have variable speed control. Some have reduction gearboxes with selection control, but the most recent system is electronic, i.e. by thyristor. A thyristor circuit may be built into the tool or be plugged in line as a separate component.

Thyristors are electronic components, 'brothers' of the transistors of modern radios. They accept alternating current and pass it on as a stream of strong impulses. By controlling the outgoing impulses the frequency has, in effect, been altered. Series-wound motor speed is dependent on the winding of the coils. If a resistor is put in circuit between the motor and the supply to cut down voltage, the motor runs more slowly; it also loses power and tends to heat up.

The resistor method of speed control is far from ideal because of the resulting loss of power and increase in heating. The thyristor method effectively cuts down the speed of supply by passing only the positive impulses of the alternating current, and these only at maximum. By controlling the supply, the motor may be run on impulses, i.e. so many to one revolution, etc. and variation in the frequency of such impulses will vary the speed of the motor. Full power is still available. Thus by increasing trigger (switch) pressure the motor can be accelerated from standstill to the maximum number of revolutions with full load all the way.

Figure 9.6 shows the principle of the series-wound motor, the low-voltage centre-earth system, the double-insulated system, the vane motor (air), and the B.S. Kitemark and double-insulated mark.

10 Sanding and Sanders

Sanding

The name *sanding* is incorrect; no sand or natural silica enters the field at all—it is a colloquialism rooted in the earliest manner of finishing wood surfaces by the abrasive action of sand. However, it is accepted, and all wood surface finishing is given the generic name *sanding*, or, to be precise, wood surface 'smoothing' in preparation for the reception of an applied surface coating (which is itself called *finishing*).

Sanding is therefore the process of fine smoothing of wood surfaces by abrasion, with the abrasive in the form of a coating on a suitable backing sheet.

The abrasives

The cutting member of the coated abrasive sheet is the 'grit' or 'abrasive'. Machine sanding uses three types of grit. They are all hard and brittle, and when fractured, leave sharp edges. All are man-made although one of them, garnet, is a natural stone which has its hardness range improved by heat treatment. The other two grits are silicon carbide and aluminium oxide, man-made combinations of silica and carbon, and molten bauxite with oxygen.

Garnet stones are mined; they are found as large crystals embedded in rock. They are hard, brittle and they fracture into evenly sized particles having sharp edges. Garnet is useful in the as-mined form, but the modern method is to heat process the natural crystal. When treated, the crystal is more uniform in consistency, it crushes into more regular grit, and has a higher hardness number

Garnet grit is red-brown in colour and is the most popular, being suitable for rough and finish sanding in hard and soft wood. Because they are tougher than the other two abrasives, they are less hard. This means that their initially sharp edges are quickly lost and the belt or sheet then settles down to a very level surface with the ability to remove wood smoothly and with a fine finish. Because of this rapid initial wear, garnet abrasives are not usually suitable for work on modern high-speed machines. They suit the lower to middle range of sander applications.

ALUMINIUM OXIDE

This is an entirely man-made abrasive. The raw material is bauxite, a soft, white rock, which is the base material for the production of aluminium. For abrasives, the powder is fused (made molten and mixed) in an electric furnace with silica and some ferrite (iron). The top surface in contact with air forms an oxide. When cold and set the resulting hard mass is fragmented.

Aluminium oxide is grey-brown in colour and is harder than garnet. Its hardness prevents rapid initial wear, and so if used on slow running machines and by hand application, tends to leave some scratches. In its smaller grit sizes, aluminium oxide is most used for veneer finishing, for it will stand up to abrasive timbers and hard glue spots. It is mostly used on power-fed sanders where constant speed and pressure are certain.

SILICON CARBIDE

Man-made silicon carbide is almost as hard as diamond. Made from silica with added coke (carbon) in an electric furnace, the resulting fusion creates a hard crystalline rock, with the right crushing characteristics. The grit particles are hard, but brittle, and tend to lose their sharp points very easily through fracture.

In use, silicon carbide has a very high cutting ability, providing that it is not too heavily loaded into the work. It is bright, shiny black in colour, and has a lustre that emery, another black abrasive used for metal working, lacks. It is rarely used for wood finishing except for some re-flooring, and in the first size-sanding of plywoods and particle boards in manufacture.

Many belts are dyed bright yellow, orange, pink or blue, and so colour is no guide to the abrasive. In general, however, coloured (i.e. not natural) belts are aluminium oxide.

The next step in manufacture is to crush and screen. Screening is size grading, the crushed mixture being shaken through progressively finer mesh screens until it is all collected and categorised into the wide range of standard sizes required. Coarse particles are for rough work and the finest for final finishing.

SCREEN SIZE

The screens used are accurately woven to give controlled openings, which are square. Passage through a screen means that in one dimension the particle is not larger than the screen size. Grit size is stated as a number, the value of which relates to the number of holes that there are in the smallest woven mesh through which the particle has passed.

In the U.K. metrication has not, as yet, caught up with abrasives grading. By international agreement, the unit of measurement for woven screen size is the *linear inch* and the number of holes such a distance encloses in the screen. A value of 120 means that the grit has passed through a mesh which has 120 holes to the linear inch. In one dimension the grit particle has a maximum dimension of 1/120 of an inch. It is probable that this grading will be with us for many years, for no metric or S.I. unit so easily fits the purpose.

The continental system has grit designated by screen size. The sizes given tend to have finer grit numbers tagged to the same size abrasive than the international system. The screen size of 100 (international) is equal in size to the continental 120.

Hand-working papers have an older, more arbitrary grading system where the scale runs from 10/0 up to $4\frac{1}{2}$, although the most used range falls between 7/0 and 2. Glass paper has its own code with such fancy names as 'double 0', 'fine two', 'middle two' and 'strong two'. These code names relate only to glass as an abrasive, and this has no machine application. However, the three machine grits do have an arbitrary coding, and some coated abrasives still carry the imprint together with the screen size. These are the 10/0 to $4\frac{1}{2}$ number referred to earlier. They run from 10/0 down to 1/0, and then rise by fractional notation through $\frac{1}{2}$ to $4\frac{1}{2}$.

Machine-used abrasives fall within a reasonably narrow range. This goes from 40 grit, being coarse, down to about 120 grit for fine work. Exceptional finish on veneered panels requires the use of belts that are even finer than this, and grit sizes may be down to 220. It is not commercially practicable to use finer grit than this. Figure 10.1 is a chart showing the comparative gradings for grit size.

Backings

These are the sheets that carry the abrasive grit. They are of paper or cloth, or a combination of both. For extra tough duty, they may be fibre reinforced and resin impregnated.

Paper backings are special and are made from jute and wood fibre, because these give fine-grain high-tensile character to the paper, which is what is required.

Cloth backings are of drills or jeans (twill), the lighter being jeans. Very fine Egyptian twill is used for flexible belts for mould sanding.

Grading chart for garnet, aluminium oxide and silicon carbide

Arbitrary code	Screen size	Description
7/0	240	Extra fine
6/0	220	
5/0	180	Fine
3/0	120	
2/0	100	Medium
$\frac{1}{2}$	60	
1	50	Coarse
$1\frac{1}{2}$	40	

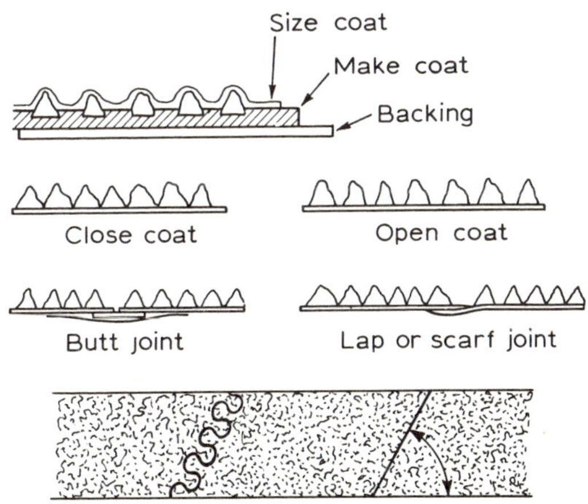

Figure 10.1 Grading chart. The table on the left gives details of a representative range of abrasive gradings, while belt castings and jointing details are shown on the right

Briefly, paper backings are for flat-belt work and pad sanders, whilst the cloth backings are used for heavy power-fed sanding on moulding sanders, drawer sanders and rough sizing of frames on drums and wide-belt machines. Where contour sanding (chair shapes, etc.) is required, cloth is necessary to withstand uneven strains.

Backings have their own code; paper is weight graded, the weight being in Imperial pounds per standard printer's ream. Continental grading for backings follows this practice of weight to area grading and the continental system is used by some English manufacturers. The system grades paper by weight in grammes per square metre. 'E' weight (machine weight) is 230 g/m^2 and 'A' weight (hand) 75 g/m^2. The 'E' or machine weight is most used, but some heavy grades of paper for rough work are of 'D' weight. Cloth backings have a similar coding with the most used being 'J', which indicates a jeans cloth. The drills have the coding 'X' which is standard for heavy applications.

Bonds

Bonding is a coating process. The prepared (glued) backing is strewn with graded abrasive particles in an even manner. The particles are embedded into the glue pre-coat or 'making' coat, as it is called, and then a further 'size' or 'bond' coat is added to lock them.

The manner of strewing produces two forms of abrasive sheet. The coating may be 'closed' or 'open'. Closed coats have particles of grit close together (more points to the unit area) than the 'open' coat which has less than three-quarters of the backing area covered. Closed coats are used on the tougher jobs of sanding, where rapid surface removal is required, and where, and this is important, the material will cut in a dry, dusty form, such as with some dense hardwoods, veneers and particle boards. Resinous (gummy) woods are not suited to closed coat work. The open spaces of open coat paper allow the collection of gummy dust without undue interference in cutting properties.

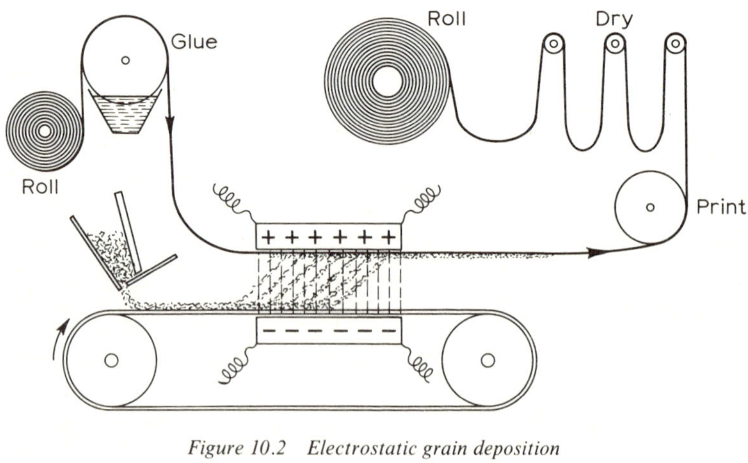

Figure 10.2 Electrostatic grain deposition

Coating is mostly done by the electrostatic process. The pre-coated backing is fed between two large platens, with the coated surface horizontal facing downwards. The lower platen is covered with a wide continuously-moving belt on which is spread an even layer of the grit that is to be deposited. The upper platen is a positive electrode, charged with a very high voltage. The lower platen is negative. Due to the exceptionally high voltage difference between the two platens, an electrostatic field is set up across the gap between them. On entering the field the loose grit particles become ionised (charged negatively) and are attracted upwards, where, having aligned their longest length directionally with the field, they impinge and stay in the glue coating on the backing. Figure 10.2 shows this process. The coated belt moves on and is given a locking coat of size. The coated abrasive, as the belt is now known, is now dried, rolled and later cut to the appropriate size.

The grit type and size and the coating type is printed, with the paper weight code, on the inner face of the belt during the drying process. A typical print-out would be:

Print-out	garnet	3/0–120	O/C	E
Meaning	garnet grit	grit size	open coat	'E' weight paper

Together with this would be the manufacturers' symbols for special types of backings and grits, and their trade imprint. When made up into belts, hand stamping would add the belt length, width and the necessary running direction.

Belt making

Belts may be made from bought-in rolls. This practice is dying out because of the superfine quality of the joints now made by belt manufacturers. Hand-made belt joints tend to be thick and lumpy.

Belt length is ascertained by calculation, or more practically, by a length of string round the running pulleys of the machine. The pulleys must be adjusted for this to allow the belt straining device to have effect on the belt, i.e. the straining device is set at something less than its maximum length. Calculation is not easy, for with sanders, belts run sometimes on two pulleys, and sometimes on three, and no two pulleys are necessarily the same size. Stringing is undoubtedly best. The belt length from the string is marked on the shop floor with bright screw heads or cold chisel markings; there is then a permanent record.

A ply cutting templet is then made. This is simply a flat board wider than the belt and with one end cut between 45° and 60°. An edge guide is added. The belt is laid, abrasive side down, on a flat board, and the templet is applied. A sharp knife or old moulder cutter is used to cut the backing material. Length is marked (along one edge) and the second end is cut to match with the first.

Belt ends are then lined up (grit down) on a piece of polythene sheet (to prevent sticking) on a flat board. Fine pins, two to each end, are driven through to hold the belt ends. Hot, thin, scotch glue is used to stick-on a narrow strip (30 mm) of thin linen. When this is smoothed down, a larger patch is added (about 100 mm) and the wet join covered with polythene. Pressure is applied by a press or weights. When dry, the protective sheet is removed and the belt is pulled up allowing the pin heads to pull through the belt. Coarse hand papering smooths the patch edges, and paraffin wax lubricates the patch. The belt should then be hung for a few hours. This is known as 'butt jointing'. Modern p.v.a. glue can be used, and for sleeve and light hand work, self-adhesive plastic tape can be used instead of patches. Layered application is made across the belt width, and then extra strips are added lengthwise along the belt.

Belts made in the above manner may run in either direction. Other shop-floor methods of belt jointing are dove-tailing and lapping.

Dove-tailed joints are made by punching the belt material down onto a sharp steel edge having an approximately dove-tailed conformation. Both ends of the belt are treated, and then, with the end tails interlocked, the standard patch method of jointing is used.

Lapping (or scarfing) produces a double thickness of backing. A narrow band of coating is removed from one end of the belt by grinding, scraping or soaking. The unscraped edge is then glued to the scraped edge and clamped. No patching is required, but the belt must run so that the lap trails into the work surface to avoid the leading edge being torn.

Joints made by the manufacturer are near perfect and it is this that has allowed development of the modern wide belt sander. The belt ends are machine lapped (scarfed) over a very short distance (9–12 mm) and pressure glued. The resulting joint is so flat that it is not obviously noticeable. However, it is lapped, and the belt should be run in the direction indicated by the arrow stamped inside. After the first long use, and to gain added sanding, most belts will stand reversal.

Sanding applications

Sanding jobs fall into three groups: flat sanding; moulding sanding and irregular contour sanding.

Flat sanding can be broken down into three classes. In the first, wood particle board, plywood and blockboard are sanded to within the manufacturer's tolerance in thickness, but on arrival at a production plant practising multi-daylight veneer pressing, all require final thicknessing for accuracy. Variation in board thickness from edge to edge can interfere with the veneering process and throw out the accuracy with which other components are fitted. Where single daylight and automatic continuous-flow pressing are used, board accuracy will not seriously effect the job. The second class of flat sanding is that of smoothing solid framework to an acceptable finish, and the third is that of final surface finishing of decorative veneers previously pressed to a base panel.

Moulding sanding can be applied to strip mouldings or to moulded edges, and contour sanding comprises of the groups of shaped carvings, chair parts, tennis racquets and all of the many three-dimensional shapes made in wood.

FLAT SANDING

Thicknessing by sanding is done on drum sanders and wide-belt sanders, each having a lockable rise and fall working bed.

Surface finishing of flat, solid components is done on pad sanders, with the edges being done on edge sanders. Final surface finishing of veneered panels is done on hand pad sanders and wide-belt sanders having floating workbeds and finishing pads within the belts.

Sanding for mouldings is not so common, for moulds themselves have tended to go out. Surface embellishments and planted panel mouldings are not suited to flow production processes, nor to application to particle-board based furniture. Those mouldings that are sanded by machine are sanded on hand pad sanders, special moulding sanders with oscillating rubbing pads and moulding sanders using automatic feeds and shaped cork blocks with which to guide flexible belts into the members of the mould.

Irregular contours are sanded on many machines; most are specially built for this class of work. There are many pneumatic drum sanders, bobbin sanders, back stand sanders, turney sanders, finger sanders and brush backed sanders. To complete the picture there are sanders that bring drawers to width and to height, sanders that sand flat ends of legs, etc., sanders that sand dowels, and those that are hand powered tools for the flattening and abrasion of lacquered surfaces.

The Machines

Pad sanders are also called *overhead-belt* sanders. They use a long endless belt of coated abrasive which runs around pulleys. One pulley acts as a driver pulley, and one of the others, if there are more than two, acts as a tensioning device, either by lead screw, weight or spring, or combinations of all three.

The belt runs horizontally above a rolling work table with the abrasive side facing down and the workpiece positioned between the abrasive face and the rolling table. Work is held against a thin stop or by vacuum cups set in the table. The belt runs clear of the work surface. A pressure pad is used to push down the belt into contact with the work and the rolling table is rocked forwards and backwards whilst the pad is progressed lengthwise along the workpiece area. In this manner the whole area of a rectangular panel is sanded. Incorporated in the headstock of the machine is a dust exhausting fan which sucks in dust-laden air and directs it along a ducting where it finally reaches a filter unit that separates dust and air, retaining the dust and venting the air. One pulley, usually an idler (not driven) acts as a tracking control. Belts may vary in length or internal tension from edge to edge. This creates a snaking roll of the belt on the pulleys. By tilting the idler pulley it is possible to centre the belt and minimise snaking. Some belts tend to run to the highest point of the pulley and some to the lowest.

Overhead-belt sanders vary mostly in their pad system. Some have no automatic or semi-automatic pad at all. The sander man holds a felt-covered wooden pad and places it on the belt. In this way he has precise control of area and pressure, and when thin veneers are being finished hand pad work is best. Special non-friction cloths that are graphite impregnated are available for pad covers.

Drum Sanders

These sand only flat panels, flat rails and flat frames. They have coated abrasive paper sheets, mostly spirally wound round metal drums previously covered with felt. Overlapping edges go into grooves cut in the drum surface. Tensioning of the paper sheet is complex. A short section of each drum (and there may be one, two, three, or occasionally eight) is spring-loaded away from main drum. The spring-loading rotates this section as it moves it away from the drum. On paper fitting, the end section is rotated and pulled firmly into the drum end. The paper, cut to parallelogram shape, is attached at one end of the drum by a steel band and then wound round the drum until its end is reached. Here the tail end of the paper is attached by a steel band to the spring-loaded section. When the paper is fully fixed the retaining pin is withdrawn and the end section rotates as it moves out, thus pulling the paper tighter. It will continue to do this and will thus keep the paper tight if it stretches in use.

The felts are either glued to the drums or spirally bound at about 50 mm pitch with fine steel wire.

A rotating drum will carry each particle of grit round in a straight line. If a panel is fed in straight feed below such a drum, and there is uneveness in the abrasive, high grits will scratch the work. Hence the drums must oscillate, i.e. move end to end. Thus no high grit will hit the work in the same straight line. Oscillation is done mechanically from one end of the drum. The bearing box is made to slide. The drum shaft extension outside the bearing carries an oscillation box. This contains the gear necessary to produce oscillation. The extreme end of the drum shaft has a worm gear cut into it that drives a small shaft by a pinion. This shaft lies front-to-back of the machine. On its end is an eccentric pin which fits into a slot in an oscillating lever which is attached to the main casting. As the drum rotates the eccentric pin rotates. When the slotted oscillation lever is engaged, the pin must stay in the same vertical line. Thus the drum is drawn back and forth to the sides of the machine. For the pre-sanding of base panels, no oscillation is used; the scratch marks give better adhesion to the veneers.

The drums are overhead, and the work, carried on an endless, serrated, rubber band, passes below. Some machines have slats of rubber-studded wood and these are called *slat beds* and are used for door and gate sanding. The bed unit is raised or lowered to suit the work thickness. It should be remembered that the province of the bed is to feed the work and to keep it pressed up to the lower faces of the drum and drum platens. To provide for accuracy in this the bed is spring-loaded and accommodates minor variations in thickness. This springing can be locked when accurate thicknessing is required.

The platen system (on a triple drum) has four platens: two between the three drums, and one in the front and one at the rear of the machine. The accuracy of the finish achieved by drum sanders is closely related to the initial setting of the drum and platens.

The first platen (all platens have flat sole plates) is readily adjustable by a hand lever, and it controls, by its height setting, the amount that the first drum removes. The higher it is above the cutting line of the drum, the more stock will be removed.

The second platen is rigidly fixed. The cutting line of the first drum must be set from this level; as panels leave the first drum they must be received and held level by the sole of the second platen.

To cut at all, drum number two must be set slightly lower than platen number two. Platen number three must be set level with drum number two. There is no operator facility for this, the settings for platens number two and three being fixed and locked by the machine maker. Drum number three is set slightly lower again than platen number three. The last platen, number four, is set in line with drum number three. It simply holds down work so that rubber feed can easily pull it out of the machine.

To summarise, this means that each drum is set in line with the platen behind it, but lower than the platen in front. Platens two and three are fixed in position

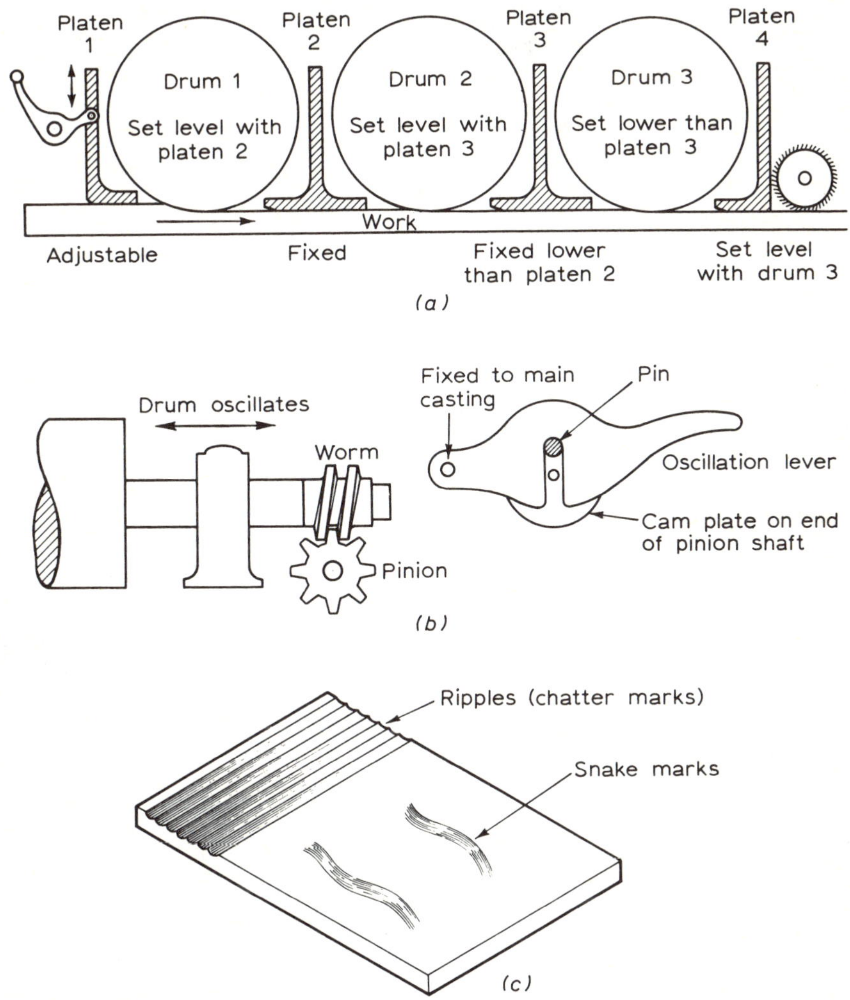

Figure 10.3 Triple-drum sander. The drum setting must be precise and to the instruction given at (a). *Drums are oscillated by the gear mechanism shown at* (b). *The chatter marks shown at* (c) *are mainly due to a poor setting relationship between drum and platens; snake marks may be due to loose paper or hard spots on the paper*

with a difference between them suited to the correct working relationship between drums one and two. Platen number one can be raised or lowered by lever to allow the operator to select the amount of stock to be removed. Work is cut down by the first drum and received by the platen behind. It is cut again by drum number two and passes to the platen behind. Drum number three gives the final finishing cut and the rear platen controls the emerging panel. A rotating brush may be fitted at the rear of the machine. All drums have fully-enclosing dust-extraction hoods. Drum setting is done by sense of touch in the first instance, using spacing blocks from the machine bed. Setting starts from the fixed platen, i.e. number two.

After preliminary setting, finish control is by handwheel. Each drum has its own ammeter (the ampere is the unit of measure of current flow). Motors react to the braking effect of work by demanding, and taking, more current, up to the stall limit (or setting of overload protection equipment). Thus a reading in amperes will indicate the power consumption and be a guide to the work being done. The heavier the cut, or the greater the rub as the abrasives wear, the higher each ammeter will read. This method gives visual information about the abrasives condition, and, more importantly, about the relative setting of the drums. Each operator must gain sufficient experience to enable him to judge the correct readings for the type of work and grades of abrasives that he is using. The trend is to use wattmeters instead of ammeters.

Snake Marks

Sanding defects from drum sanders mostly become either obvious ripples across the work or else they show as wide-band snake marks down the length of the work. Ripples are a sign of poor bearings, unbalanced drums, feeds that are too fast for the depth of cut, or that one or more of the drums are not in line with their following platens.

Snake marks have many causes. They usually indicate a fault in, or on, the abrasive sheet. Glue spots, resin build-up, loose spots due to damp penetration during store, and incorrect application (tensioning) are the main causes. Other reasons for snake marks are loose felts, dust pockets, hard felts and lumps on the feed belt.

Figure 10.3 shows details of a triple drum sander.

WIDE-BELT SANDERS

These are so numerous that only a representative machine can be included here. Wide-belt sanders, as their name implies, use wide belts of coated abrasives. The immediate advantage of this over drum sanders lies in the extra length of abrasive. Drum sanders are restricted in length to the drum circumference; wide belts have as long a length as the machine permits.

Design principles are as follows: a wide belt (or belts, on a larger machine) runs around an idler roller which acts as a tensioning and tracking control. At the work surface the belt runs on a contact roll which is a rubber-faced pulley having a rubber hardness code suitable for the general run of work. A helical, square groove is cut deeply into the rubber. As the groove runs around it leaves a 'land' between each turn approximately as wide as the groove itself. In effect, half of the roller surface is cut away. Various feed rolls or endless rubber belts feed and control the passage of work panels under or over this roll which carries the sanding belt. The machine is (so far) a 'contact roll sander'. The next stage is an added steel roller spaced out behind the contact roll. Between the steel roller and the rubber contact roll, a narrow, resilient pad is inserted. This, in use, lies below the line of the contact roll and presses the belt down (or up) onto the work. Work fed through is cut down to a reasonable finish by the contact roll and finally smoothed when the pad gives a sensitive pressure contact. The grooves allow a shearing form of contact, as only half the belt surface is pressurised at once. The belt, unsupported by the space between lands, sags, shakes free its dust and loses any heat build-up. Thus the contact roll system gives a longer life to the coated abrasive and a better finish from a coarser grit. The pressure pad is foam faced and is arranged to lift away as the tail end of the panel passes, and to lower only after the leading end of the panel has passed. This prevents 'dubbed' or 'turned-over' ends.

This is the single belt machine. The variations are mainly in the number of belts and the combinations of contact roll and pressure pads. Most machines have separate belts for contact roll and pads, some have two rolls and one pad, others have belts top and bottom; the full range is wide.

Belt Control

All belts tend to ride sideways on their pulleys. This must be controlled as oscillation is not necessary with contact roll applications. Belt oscillation is not a serious defect as it tends to improve surface, but it must be controlled. Control is by movement of the tensioning pulley. Many systems are in use to tilt or pivot the tensioning pulley in order to track the belt. Some belts run uphill, some down, so tracking means that the run of the belt must be detected and then opposed by induced opposite movement.

Nearly all systems use a compressed air cylinder to raise or lower the pulley centres as necessary. The systems differ in the sensing circuit used. One system uses compressed air which is fed into the roller which is divided into halves. Each half is vented with micro-bore holes at the end of the outer surface. As the belt runs to one end of the pulley it covers some holes and opens others. Sensitive air-control valves sense this and move the pulley end to reverse the belt drift.

Another system has, at either side of the belt, a low-pressure air jet and receiver. The jet blows into the receiver and its pressure is magnified sufficiently to operate the control valve of the tracking cylinder. As the belt drifts it cuts one jet of air and opens the other at the other side. Thus it is directed back and forth across the pulley.

Other systems have air jets keeping micro-electric switches closed. As the belt drifts the switch is opened and a solenoid valve controlling the air flow to the tracking cylinder is also opened. The repeat system at the other side of the belt then moves the cylinder to pull the belt back.

Tensioning of belts is by a balanced compressed-air system. Air pressure drives a main pulley out to tension the belt, but lower pressure air is fed in to compensate for tension and to act as a cushion. Should the air pressure fall significantly, the electricity supply is cut off, and on some machines the brake is applied. All machines are fitted with brakes to protect the rubber feed and contact roller should the belt break or over-run to the side. Various methods of detecting over-run are used. One machine has a thin elastic band holding-in a micro-switch. If the band is cut by the belt, the switch cuts all power supply and applies the brake through a relay.

Use of Wide-Belt Sanders

Contact roll sanding is a stock removal process. The hard rubber, together with its grooved formation, allows heavy abrasives to be carried, and this enables rapid and accurate removal of surface. When fitted with fine cutting belts the contact roll is able to give an acceptable finish, but not one that is as fine as that given by fine belts operating through a pressure pad. These pads cause longer belt contact than do contact rolls, and thus produce flatter surfaces. For thicknessing only, the pad is raised and the table is locked. For fine finishing the pad is brought into use and the table is floated (if this facility is incorporated).

Power, speeds and grades

Belt speeds (velocities) are measured in metres per minute (m/min) and power in terms of watts and kilowatts (W and kW). Grades are given in screen size (holes per linear inch).

SPEEDS

Resin-bonded and linen-backed belts may run faster than glue-bonded paper belts:

Glue belts	900 m/min
Resin belts	2100 m/min
Wide belts for wood	920 m/min 1520 m/min
Drum speeds	1000 m/min to 1200 m/min
Pad (belt) sanders	1000 m/min to 1200 m/min

FEED SPEEDS

Wide belts	10 m/min to 30 m/min
Multi-drums	6 m/min to 18 m/min

Note: all of the above speeds are given as a reasonable average. Belt condition, work type and machine condition all affect possible working speeds, as do abrasive type and size in use and the surface finish required.

POWER

Power used is proportional to belt width, cut for cut, but so many factors affect this that power rating can only be generalised. Coarse grits require more power

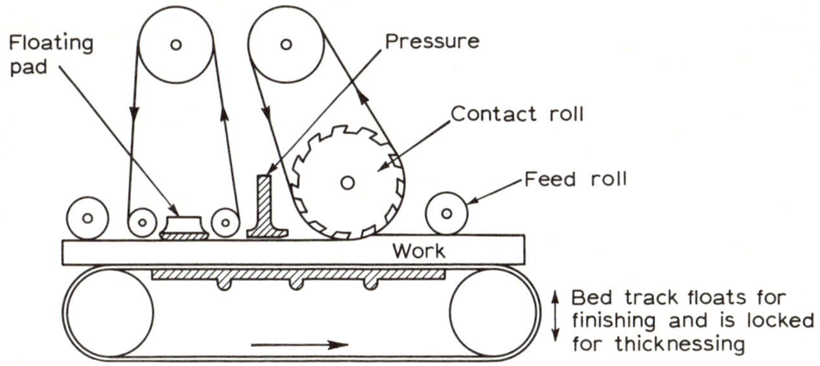

(a)

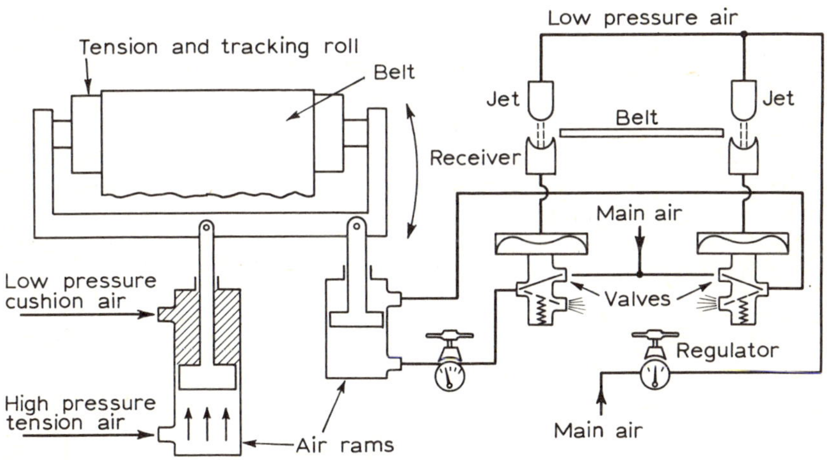

(b)

Figure 10.4 Twin-head wide-belt sanders are generally arranged as shown at (a). *Belt tracking is usually by compressed air, and* (b) *shows the basis of a typical system. Low-pressure air jets are magnified by diaphragms that allow air to flow through valves. Either jet interrupted allows the piston to move up or down accordingly*

that fine grits. Soft contact rolls require less than hard ones. High feed speeds use greater power than slower speeds.

WIDE-BELT MACHINES

The power for these is usually approximately proportional to the belt width.

600 mm width 7 460 kW
1125 mm width 18 650 kW

DRUM SANDERS

The third drum of a triple drum machine often has a slightly lower-powered motor than the first two drums.

630 mm width 746 kW
1825 mm width 18 650 kW

BELT SANDERS (PAD)

150 mm width 4 213 kW

Sanding grits

Grade selection depends on the surface finish required and the nature of the material. On average, wide-belt machines use from 80 to 120 for reasonable finishing to size, but up to 200 for veneer panels when the contact pad is in use.

Triple drum sanders tend to be fitted with one of two grade chains: for reasonable work 40–50–100, and for panel work 50–80–120. Belt (or pad) sanders use belts to the operator's choice. Softwood painted work is acceptable from 80 grit belts or lower. Veneered panels require belts up to 150, but remember that garnet tends to give a finer finish than aluminium oxide, grade for grade, but cuts more slowly and wears more quickly.

Figure 10.4 shows a section through a typical two-head wide-belt sander and includes a representative belt tracking layout.

11 Dowel Boring

Strength

Dowel jointing has been traditional for chair making since it was first found possible to bore accurate holes in wood. Terminology is loose, but holes made to receive dowels are usually 'bored', not 'drilled'—a difference in trade vernacular but not in actuality.

The accuracy which it is now possible to work to with today's machines and drilling bits (drills, bits, wood borers, etc.—they all make holes in wood) is such that many jobs other than chair construction are dowelled, and with adequate strength for the purpose. Dowels are often the most suitable form of edge jointing for particle boards. The strength of a dowelled joint is relative (among other things) to the fit of the dowel in the holes, with the fit being of most importance. Dowel fit is another way of saying 'glue line thickness', for as the dowel gets closer in diameter to the size of the hole so the glue line gets thinner, and from this comes the corollary that as glue lines get thicker, joints get weaker.

The bits

Dowel bits vary. The choice lies between high-carbon steel bits of 'Jennings' pattern, with either a centre screw point, or brad, and the type known as the 'machine dowel bit', which has a centre brad point, two cutting lips and two spurs. The brad projects about 1·5 mm beyond the spurs, which again project about 1·5 mm beyond the lips. These bits are often called *lip* and *spur* bits. They are much shorter than normal for Jennings bits and can be bought in high-carbon steel, h.s.s. or tipped with tungsten carbide.

The helix angle of the clearance flute is much less than that of the Jennings and this assists the chips away more quickly. One form of lip and spur bit has one of the 'lands' between the flutes ground away leaving a solid centre, as have some mortise augers, and for the same reason, i.e. chip clearance. At all times the centre brad has the job of providing an accurate turning point for the bit without which it may wander along grain lines.

Dowelling is not very successful if twist drills are used, for it is difficult to maintain the dead accuracy of the point, and without spurs the grain is not so cleanly cut. Twist drills do have a use in dowelling—if the dowels are oversize and no dowel sander is available, larger holes than the nominal drill size may be made if the drill point is pushed off centre by grinding only one leading edge. The drill then rotates round its new centre and makes a larger hole. However, this is an expedient and is not too reliable, for as timber density and moisture content varies and as feed speeds vary, the size of the hole will vary, with force feeding in hard material producing the largest hole.

Boring practice

Bits must have sharp cutters and sharp spurs, and the centre brad must be sharp and central. For flat work, either in side or end grain, the spurs should closely follow the brad, but for angular penetration of curved surfaces the spurs must

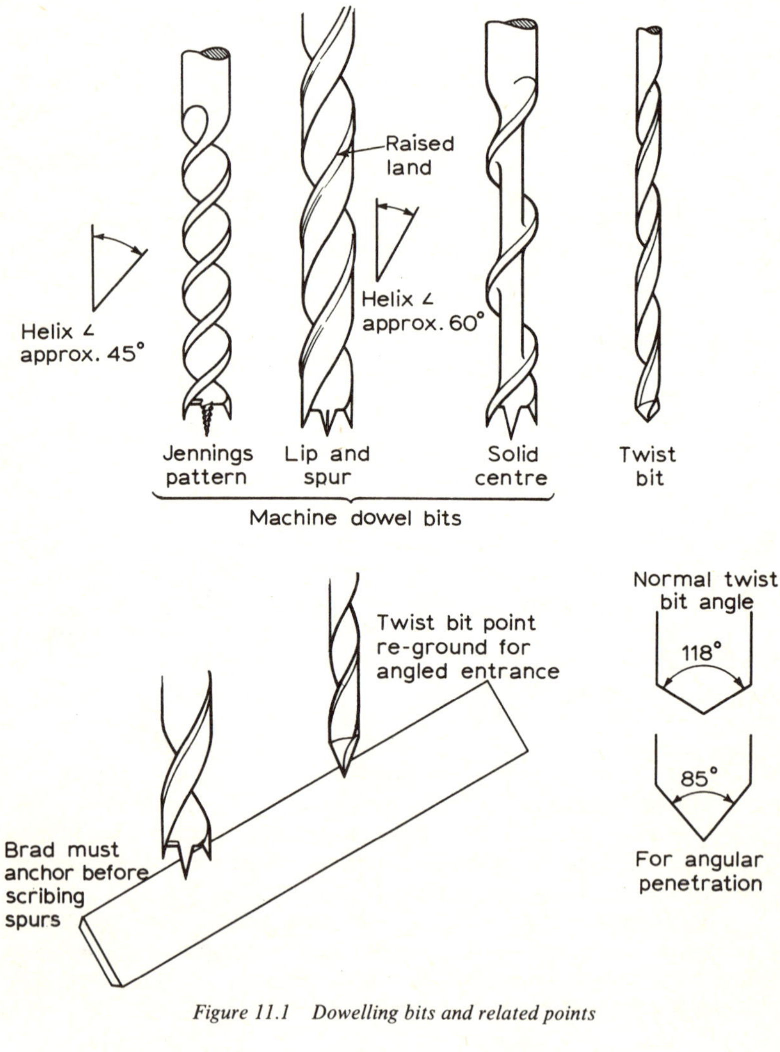

Figure 11.1 Dowelling bits and related points

be set back sufficiently to allow the brad to centre the bit before the spurs start to cut. This is done by very careful grinding and accurate filing (*see* Bit Maintenance).

To be accurate, holes must be cut with sharp bits that are forcibly fed and that are retracted-out immediately. Burn-polished wood never glues as well as cleanly cut wood, and this applies equally to dowel holes. The insides of the holes must appear smooth, not woolly, for woolliness indicates fibre breakdown and gluing to broken fibres produces weak joints. For extreme angled penetration, twist drills having a very low enclosed angle of point are sometimes better than spur bits, but suffer from the defect of wasted depth and woolly entry.

Figure 11.1 shows a range of dowel bits and visually explains some of the above points.

The machines

Dowel borers range from single horizontals to multiples having both horizontal and vertical heads, all with angular setting facilities and with additional angular cut-off saws.

Controls range from hand loading and hand feeding to machines on fully automatic cycle for load, clamp, bore and eject. Most recent are the tape, plug board, or punched-card-controlled machines that have end loading and built-in self-setting features for the heads.

SINGLE-DOWEL BORERS

The name *single* is not fully apt, for although there is a single boring head it may carry as many as five dowelling bits in single clusters, or up to four spread out in line.

The basic features are always the same. There is the head unit, which moves forward carrying the bits into the work (a few machines carry work into the bits) and which may carry single bits or multiple clusters driven from one drive unit. As the machine design progresses this head may be arranged to cant and to pivot for angular boring, although most angular boring is done by jigged presentation of the workpiece.

In front of the sliding head there is a work table which carries an adjustable fence and end stops, together with a work clamp. This table may be raised or lowered for position setting. It may also tilt from front to back, or, occasionally, side to side, although this facility is usually given by attachments.

The work process is to set the machine cluster with the appropriate bits and for the head stroke to be set for depth and speed. Table fence and stops are set, together with the air clamp. Table angle is also set. The workpiece is then positioned and the foot pedal depressed. Some machines operate mechanically through linkage, but most modern machines have combinations of applied pneumatics and hydraulics. A popular method is to use self-feeding drills, either having air motors or electric motors, with forward feed being by air.

The normal unit uses an electric motor to run bits and an air cylinder for feed and return, with the precise speed being controlled by oil flow through restrictor valves. These combinations are variations of hydro-check units where oil actually creates the effective forward thrust, but is itself thrust, or pressurised, by compressed air.

Whatever the system, once the foot pedal is depressed the clamp locks the work, the bit cuts the hole and retracts, and the workpiece is released, with exhausting air from the clamp being used to blow chips away. For shaped work (curved rails) various saddles are necessary, and these are bolted to the machine table. The table pivot is indexed in degrees.

Feed-in speed has a variable control through a handwheel which opens or closes the oil-control flow valve. Feed is controllable from a very fine minimum to a fast maximum. The flow valves allow full flow to the return oil and the return stroke is always at top speed.

Hole depth has two controls, i.e. the position of the work fence relative to the edge of the table, and the setting of a handscrew stop on the feed slide.

Feeds should be as fast as the resulting hole will allow. All drills (wood or metal) need a heavy feed to cause the restricted cutting edge to turn back the chip and fracture it.

The setting technique is to mark accurately a work sample and then to operate the machine head until the point at which the in-feed ceases (hole depth) is found. The workpiece is aligned on this line. The feed-speed control valve is screwed in until it allows the minimum in-feed speed. Bits are brought forward until they overhang the table and the head screw is fully tightened to lock movement. By manipulation of the cluster collar, table height and tilt control, and by the positioning of stops and work fence, all bits may be set to their correct marked centre.

Depth is also adjusted. A running check will show an inaccuracy that requires final adjustment.

The Cluster Heads

The two main types of cluster heads are the completely round self-contained unit which fits into the machine nose and which carries bit chucks at set centres, and the older type that has each bit driven from its own shaft which has independent adjustment up or down or across the machine. These are usually limited to four and are restricted for close centre work. They are, however, useful for spreads of holes in approximately straight lines such as may be used for wide table rails.

Combined clusters are driven from a central toothed shaft that fits centrally to the individual shafts of the bit holders. All the gears are contained within the cluster unit and are fully oil bathed. Clusters may be interchanged to give different centres.

Most dowel borers have bits that have screwed shanks. One or two machines may have standard drill chucks fitted for larger holes.

The round cluster units may have a very wide array of hole centres. A five-bit unit may have centres set to 15, 17·5, 22, 24 and 30·5 mm. One, or all five bits

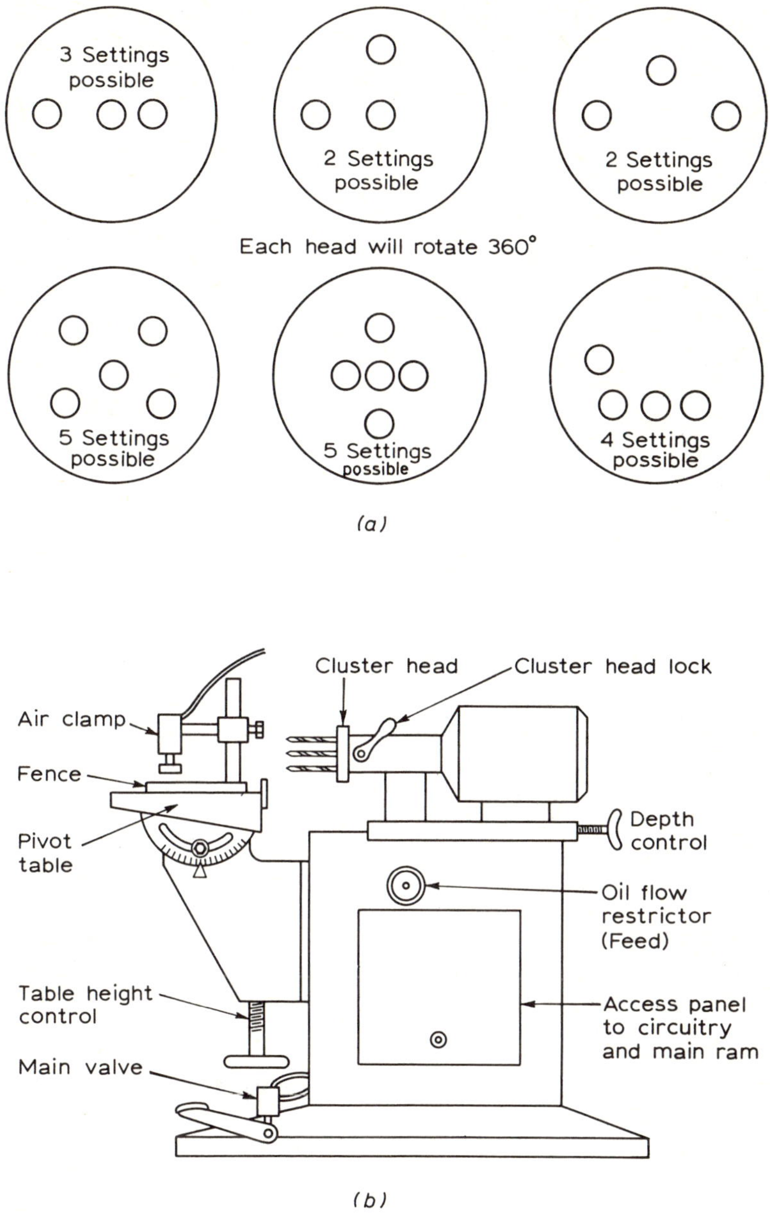

Figure 11.2 Representative range of cluster heads using two bits are shown at (a). *The main features of a single-head dowel borer are shown at* (b)

may be used, or combinations of them. Four bit clusters have centres of approximately (according to the maker) 17, 24, 27 and 34 mm. Some have the bit centres set evenly round a central bit and by selection various bit patterns may be used. One such cluster (Brookman) has standard centres of 15·8, 19, 22·2 and 25·4 mm.

The cluster unit itself will rotate through 360° in its holder which enables almost unlimited selection of dowel hole pattern, especially if the head will take five bits. Cluster units and bit patterns are shown in Figure 11.2 together with a general outline of a single-head dowel borer.

TWIN MULTI-HEAD DOWEL BORERS

Some single-head machines are used as a base for twin heads. This extends the range and opens out the maximum centre distance possible. Nearly all heads on single machines move forward on machined tracks, and thus can only bore holes that align front to back of the machine.

On larger multiples, heads are often of the self-feed type where each unit has its own drive motor and self-contained feed unit within its own casing. Thus angular tilt and pivot is possible for each bit will bore along the line on which the head is angled. These multi-head machines also have the multi-bit clusters of the single machines. Heads are arranged horizontally and vertically, and on larger machines there are both, and, in addition, two compound-angle cut-off saws. All heads can be controlled by one automatic cycle from foot pedal initiation. With heads pivoted on the base through 90° to face each other, rails may be cut to length with compound angle, dowel bored on ends and side faces, and ejected from the machine.

Spigot-cutting units (round tenons or solid dowels) may be fitted to some heads of the heavier machines whereby rails may be length trimmed and produced having angled or parallel spigots.

THE 'DEXTER' MULTI-SPINDLE MACHINE

This is a multi-compound angle borer. It will take heavier bits than most universal borers and has an enormous facility for angular setting in the minimum amount of time. Each bit is driven by a flexible drive shaft from a unit motor, and each motor will drive up to six bits. As many units can be assembled as are required.

The system is that each bit will have its own setting block, or jig, which will present it to the work at the required angle. These blocks are slim and close fitting to permit short centre boring. Once a boring scheme is planned, each bit has its own jig made and this is attached by one bolt. The jig raises the bit shaft for height and tilts it to the angle required. Thus if a bit must bore a hole 20 mm out from its neighbour and at 15° to it, the jig provides for this.

Forward drive (into the work) of each bit is by sleeved cable and each cable is connected at its other end to a control unit. When the main beam moves, each bit shaft connected by cable moves. Selection simply means connection of the appropriate cable. The flexible drive shafts to each bit are driven by timing belts from one main pulley. The use of flexible drive shafts and sleeved cables for in-feed means that each bit is an individual unit and may be set without regard for the other five units in its team.

A suitable application is that of boring the angled holes of solid Windsor chair seats.

Figure 11.3 shows outlines of various boring and drilling machines, while Figure 11.4 gives details of the 'Dexter'.

Bit maintenance

Bits are filed to sharpness with slim fine-cutting warding files or swiss files. With skill and the correct wheels, initial sharpening can be done by grinding.

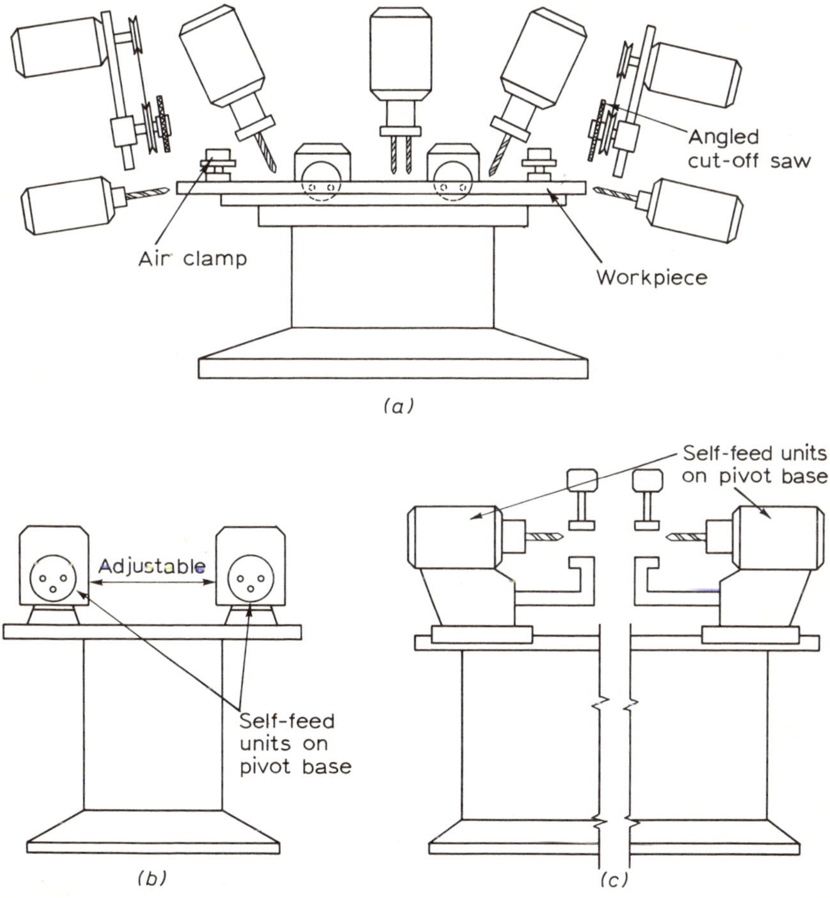

Figure 11.3 *Range of dowel borers.* (a) *Multi-head cut-off and borer* (*all heads sequenced from pedal control*), (b) *twin head* (*sequence and angles adjustable*) and (c) *double-ended borer* (*up to 2 m*)

Machine dowel bits are filed on the brad point, the two lips and the inner sides of the two spurs. Flutes must be polished and kept clean. Long bits must be rolled on a flat plate to detect bends. If a bit is bent it is not much use, although some correction is possible using sharp taps with a hardwood edge on the correct land.

For sharpening, the bit is held wrapped in a rag with the point up in the vice, and the brad point is filed to a sharp central point. Then the inner (or upper) faces of the cutting lips are filed; finally the inner faces of the two scribing spurs are filed. At all times the file must be fine toothed and must remove no more than is necessary to bring up a slight burr on the outer face of each cutting part

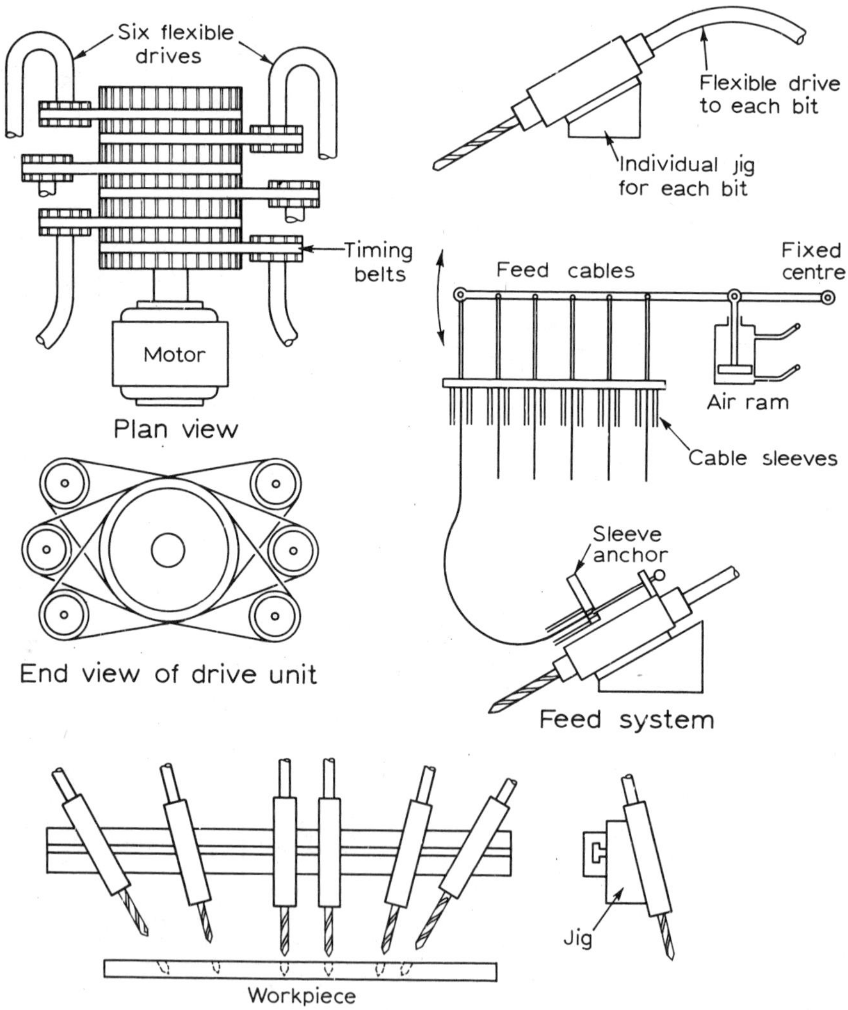

Figure 11.4 The Dexter Multi-Borer is a 'special' and all its components, in units of six, can be built into any structure

sharpened. Lips should be maintained about 1·5 mm behind the spurs, with the brad point out in front.

Power and speed

Cluster units require about 373 W per drilling spindle, but because this would be catering for maximum load at all times, motors of only 1492 W are normal. Bit speeds range from 2200 to 3000 rev/min, with the higher speed probably cutting more accurate dowel holes, but this depends so much on bit sharpening, etc. that it is not a hard and fast rule.

12 Technical Data

All technology derived from and based on craft skills, must include a wide range of fringe area subjects. Machine woodworking is no exception.

The man in charge, whether he is joiner, cabinet maker or machine woodworker, must have a knowledge of tools, processes, drive methods, machine design and a background knowledge of cutting theory, and so on. The knowledge may not be categorised or tabulated, but it must be there as a reservoir to be tapped as required. This chapter is intended to cover some of these fringe area subjects.

Machine speed calculations

Most examination papers that include questions on machines include calculations, and these mainly relate to belt driven operations. Belt drive calculations have four factors: drive unit (motor) revolutions per minute, drive unit pulley diameter, machine shaft speed (rev/min) and the machine pulley diameter. Leaving out belt slips, these four factors are related as follows.

The motor runs at a given number of revolutions per minute and drives its pulley at this speed. The surface (or peripheral) speed of the pulley is given by its number of revolutions per minute multiplied by its circumference. The belt velocity is equal to the peripheral speed of the pulley. At the other end of the drive, the machine pulley will be driven at the same peripheral speed. The machine pulley may be larger or smaller than the motor pulley, and will run slower or faster, accordingly, in terms of the number of revolutions per minute.

THE CALCULATIONS

Belt velocity V (m/min) is given by the formula

$$V = \frac{\pi D R}{1000}$$

Where D = pulley diameter (mm),

R = number of revolutions per minute (rev/min),

π = 3·14 (ratio of pulley diameter to pulley circumference) and 1000 is the divisor to convert millimetres per minute to metres per minute.

Having found belt velocity by using this formula it is necessary to divide it by the circumference of the driven pulley to give r which is the number of revolutions per minute of the driven pulley, which is numerically equal to the speed of the machine. Thus machine speed (rev/min) is given by the formula

$$r = \frac{\pi DR}{1000} \div \frac{\pi d}{1000}$$

Simplifying
$$r = \frac{\pi DR}{1000} \times \frac{1000}{\pi d} = \frac{DR}{d}$$

Where D = driving pulley diameter (mm),
R = driving pulley revolutions per minute,
d = driven pulley diameter (mm) and
r = driven pulley revolutions per minute.

By transposition the four equations that are possible are:

$$D = \frac{dr}{R}, \quad R = \frac{dr}{D}, \quad r = \frac{DR}{d} \quad \text{and} \quad d = \frac{DR}{r}$$

Other symbols may be used, but these fit best. Remember that capital letters are used for drive factors and small letters for driven factors.

Countershaft drives have a third series of factors. The motor drives a pulley, that drives a belt, that drives a pulley, that drives a shaft, that drives a pulley, that drives a belt, that drives the pulley that drives the machine. The countershaft carries the two intermediate pulleys, and these can be of different diameters, thus giving differential belt velocities. Problems involving countershafts have the same method of solution. Consider the countershaft as the machine shaft. Then countershaft revolutions per minute become

$$r = \frac{DR}{d}$$

This becomes the driving speed of the second part of the drive, and the full calculation becomes

$$r = \frac{DPR}{dp}$$

Where P = countershaft driving pulley diameter and
p = countershaft driven pulley diameter.

Example 12.1

Find the number of revolutions per minute of the machine where the motor runs at 3000 rev/min, and has a pulley of 250 mm diameter driving a countershaft pulley of 200 mm diameter. The second (driving) pulley on the countershaft is 300 mm diameter, and this drives a machine pulley of 150 mm diameter.

The formula is
$$r = \frac{DPR}{dp}$$

substituting
$$r = \frac{250 \times 300 \times 3000}{150 \times 200} = 7500 \text{ rev/min.}$$

By rearrangement of the above factors, any missing factor may be found if all the others are present.

$r = DR/p$ which is the number of revolutions per minute of the countershaft, so the formula DR/p may be used as a self-contained symbol and be combined with other symbols to give a formula. It is not necessary to calculate the number of revolutions per minute of the countershaft. Always compile the complete problem, then cancel, then multiply through.

BELT SLIP

All normal belts slip, or creep. Flat wide belts do so the most, 'V'-belts less so, flat nylon and pigskin belts less still, and timing belts not at all. Wide balata belts (fabric and rubber) and leather belts slip at least 5%. This means that the number of revolutions per minute delivered at the machine are 5 in every 100 less than at the drive end (5% = 5/100). Belt slip is dealt with by subtracting the percentage rate from 100 and then using the difference as a multiplier. It is not necessary to calculate the percentage of a given number and then to subtract this from the total, i.e. 5% = 5/100, subtracting leaves 95/100, so multiply by this.

Example 12.2

The speed of a motor is 3000 rev/min, the pulley diameter is 150 mm, the machine pulley diameter is 100 mm and the slip is 8%. Find the number of revolutions per minute of the machine.

$$r = \frac{DR}{d} - 8\% = \frac{92}{100} \times \frac{DR}{d} = \frac{150 \times 3000 \times 92}{100 \times 100}$$

Hence $r = 4140$ rev/min.

PULLEY SIZE CALCULATIONS

Given a peripheral speed, what size pulley is required either at the motor or the machine shaft to achieve it? This type of problem involves π.

Example 12.3

A motor running at 1500 rev/min has a pulley of 250 mm diameter. To cut reasonably well the cutters must run at 1500 m/min. What diameter pulley is needed on the block shaft? The block diameter is 125 mm.

Solution

Given the block rim speed in metres per minute, the required number of revolutions per minute may be found by dividing this given speed by the cutter-block circumference. To find the machine pulley size, the machine speed (rev/min) must be divided into the product of motor revolutions per minute times motor pulley diameter.

$$d = \frac{DR}{r}$$

where d = machine pulley diameter,
r = machine speed (rev/min),
D = motor pulley diameter and
R = motor speed (rev/min).

Then block pulley speed r (rev/min) is given by the formula

$$r = \frac{1000P}{\pi D_B}$$

Where P = stated rim speed (m/min),
π = 3·14 and
D_B = block diameter (mm).

Combining the two formulae and rearranging gives

$$d = \frac{\pi D_B DR}{1000P}$$

Substituting, this formula gives

$$d = \frac{3{\cdot}14 \times 125 \times 250 \times 1500}{1000 \times 1500} = 98{\cdot}125 \text{ mm}$$

or, to be practical, a pulley of 100 mm must be fitted.

Balance in cutter blocks

The theory of cutter-block balance was explained in Chapter 6. The calculations relating to this are more complex than those relating to machine speeds, and they come under the heading 'centripetal force'.

CENTRIPETAL FORCE

The 'Law of Inertia' states that a moving body will continue to move in a straight path until a force is applied to change its direction of movement. Inertia can be said to be the resistance to movement of a body.

A cutter bolted to a rotating cutter block must, by necessity, rotate. Its desire to fly off at a tangent is thwarted by the pull of the nut and washer which act as a force to bend or distort the cutter from its desire to fly off in a straight flight. The force applied by the nut and washer is called *centripetal* force, and was named this by Newton, and meaning in Latin to flee to the centre. This is not an easy concept, yet it is one that is true. All parts of any moving body are subject to acceleration; they are at rest, and then they move, i.e. they are accelerated. As they accelerate they 'desire' to go straight but are not permitted to do so, i.e. they are bent out of this straight flight by the pull of the centre (centripetal force). This difficulty of mental appreciation was quickly seen by the early philosophers and mathematicians, and they therefore devised a reasonable alternative. The Latin name chosen means to flee from the centre, and the force was called *centrifugal* force. In all respects centripetal force may be forgotten. In order to remain in 'balance', all forces must be equal and opposite and thus all that relates to centripetal force may be 'turned on its head' and related to centrifugal force, the outward pulling force of rotating bodies. The calculations are the same, only the name is changed.

The Calculations

The formula used for centrifugal force calculations depends on how far back to the basic principles one wishes to go. The roots of the problem lie in g the accelerative force due to gravity, ω the velocity ratio and in acceleration (V^2/R). However, most problems of this sort for woodworking students are solved by using a short formula containing a constant. This is itself derived from another standard formula, and the process is as follows:

$$F_C = \frac{MR(2\pi N)^2 \times 10}{g \times 60^2}$$

Where M = mass (kg),
R = radius of gyration of approximate centre of gravity of M (m),
N = number of revolutions per minute,
F_C = centrifugal force (N),
g = 10 kg (approximated from 9·807) and 10 is the multiplier to change kilograms to newtons.

The mass is the weight of the out of balance part of the cutter, i.e. the amount by which one cutter is heavier than the other; g is the accelerative force due to gravity. The formula is simplified into variables and a constant as follows:

$$F_C = \frac{MR(2\pi N)^2 \times 10}{g \times 60^2} = \frac{MR \times 4\pi^2 N^2 \times 10}{10 \times 60^2} = 0{\cdot}01MRN^2 \text{ (newtons)}$$

This numerical coefficient will always be constant, and the only variables will be M, R and N.

Using the Imperial g of 32·2 lb the formula derived will be

$$F_C = \frac{WR(2\pi N)^2}{g \times 60^2} = \frac{WRN^2 \times 4\pi^2}{32{\cdot}2 \times 60 \times 60} = WRN^2 \times 0{\cdot}00034 \text{ lb}$$

and this is the Imperial formula for centrifugal force (lb), where W is the weight (lb), R is the radius of gyration of unbalance (ft) and N is the number of revolutions per minute.

Using S.I. units it will be sufficient to solve MRN^2 and then to move the decimal point two places to the left. The mass must be in grams and the radius in millimetres.

Example 12.4

Calculate the out of balance force when a mass of 14 g (0·5 oz) rotates at a radius of 75 mm (3 in) at a speed of 4200 rev/min.

S.I.	*Imperial*
F_C (in newtons) $= 0{\cdot}01\ MRN^2$	F_C (in lb) $= WRN^2 \times 0{\cdot}00034$

where M and W = mass (kg) and weight (lb) of unbalance,
R = radius of gyration of unbalance and
N = number of revolutions per minute.

Substituting

$$F_C = \frac{0{\cdot}01 \times 14 \times 75 \times 4200 \times 4200}{1000 \times 1000} \quad \text{and} \quad \frac{1 \times 3 \times 4200 \times 4200 \times 0{\cdot}00034}{32 \times 12}$$

$$= 185\ \text{N (or 18·5 kg)} \quad \text{and} \quad 46{\cdot}85\ \text{lb (or 21·3 kg)}.$$

Allowing for the differences between 75 mm and 3 in, 14 g and 0·5 oz, and the approximation of 9·81 N as 10 N, the answer is close enough to demonstrate the method and the formula.

Grinding

To grind is to abrade, i.e. to strike continuously for the purpose of removal of material. In machine woodworking this means removal of metal by grinding.

Grinding wheels are a matrix of abrasive particles held together with a bonding material. Each particle of abrasive strikes the metal being ground until one of two things happens. Either the sharp cutting edge of the abrasive particle will wear until it will cut no longer (at this stage the particle must fracture to reveal a new sharp edge), or, the load on the particle must be sufficiently high to forcibly remove it from the embrace of the bond. The bond material must then be abraded away by the metal being ground until fresh particles of abrasive are brought into use. The process is continuous, and under ideal conditions, grinding wheels are said to be 'self-sharpening'.

The factors involved include the type and size of the abrasive grit, the nature of the bond material, the speed of the grinding wheel (peripheral velocity) and the nature of the material being ground. The factors relate in the following way: a hard, brittle metal-like material (tungsten carbide) shatters on high impact and heats rapidly when subjected to rubbing friction. Therefore, to cut it smoothly requires very sharp edges on very hard abrasive grit. To do this, the grit must be held at normal cutting force, but must not be held so tightly that it will not fall away when its sharp edge is dull. This may be summarised as 'the harder the material being ground, the softer must be the bond of the wheel', relatively speaking, of course. The corollary of this is that when softer materials are being ground that tend to drag rather than be brittle enough to fracture away, much harder wheel bonds are necessary. The softer the material being ground, the harder the wheel bond must be.

GRITS

The grinding grits (abrasive particles) in use for machine woodworking tools are aluminium oxide [a combination of aluminium and oxygen (*A*)], silicon carbide [silicon and carbon (*C*)] and diamonds [industrial diamond (*D*)]. Aluminium oxide is available in two forms: standard and white. Silicon carbide also has two forms: standard and green.

Each grit has its own range of work. Diamond grit is the hardest and is used for the finish (or lap) grinding of tungsten-carbide tool tips. Silicon carbide (green grit) is used for the rough grinding of tungsten carbide, while standard silicon carbide is not used at all on woodworking cutters. White aluminium oxide is used for h.s.s. cutters, while standard aluminium oxide is used on carbon-steel tools.

Abrasive particles are graded for grit size by screening. Screen numbers still remain Imperial at the time of writing, and indicate the number of holes there are per linear inch in the screen. Grit size 80 indicates that each particle has passed through a hole of that size (1/80th of an inch). Woodworking cutters require grit sizes ranging from 36 up to 80 for normal tools and up to 220 for diamond wheels that lap (polish) carbide.

BONDS

Bond materials suit the conditions required by the grinding process. Hard, brittle metals require softer bonds than softer, tougher metals. The bond hardness is called the *grade* of the wheel. The materials used are vitreous clay, silicate of soda, rubber, metal (bronze type), resin and shellac.

Of these, vitreous clay is used for general purpose grinding, and is called 'vitrified bond' (*V*). Silicate of soda (*S*) is for very special thin edge tools, and is not used much in general woodworking. Rubber (*R*) is used for very thin wheels that are suitable for use as slitting wheels to cut off wide bandsaw steel. Resin or resinoid bonds (*B*) are found on trough grinders and other wet grinders, usually for h.s.s. grinding; they are sometimes used for thin wheels, and for some cup-shaped diamond wheels. Shellac (*E*) is not a woodworking bond. Metallic bonds (*M*) are sometimes used to hold diamond grit for carbide-tipped tools.

SPEED

The speed of the wheels affects the load on the abrasive and bond. High speed gives high impact and easy metal removal; low speed gives such impact that the load approaches the ultimate strength of the bond and grit, and thus induces rapid wheel wear.

A wheel operating at one speed will change its character if operated at another. Thus wheels appear to be softer if run more slowly. In practice this means that once a wheel has worn to a smaller diameter it will begin to wear much more rapidly. Conversely, if a wheel is run at too high a speed for its grade it will not clear itself, and will burn the tool.

Wheels are stamped with the maximum revolutions permitted, and under the Abrasive Wheels Regulations (1970) each machine must have its mandrel speed clearly marked.

CODING

Information relating to wheel characteristics such that it will affect selection is coded and carried on each wheel. A typical coding is

	WA	*80*	*K*	*V*
which means	Abrasive type	Grit size	Bond hardness (grade)	Bond type

This would mean a wheel of white aluminium oxide of 80 grit, having a bond hardness of *K* (medium) and a vitrified type of bond. This would be suited for grinding h.s.s. solid-profile cutters.

The hardness symbols run from *D* to *K* with only the middle range for woodworking. *D* is the softer end as coded by British Standards, but some manufacturers reverse this code.

WHEEL SHAPES

Flat wheels (discs) are used only edge on; there are a variety of other shapes that are suited for special work.

Cup wheels are cups that have parallel walls and are used on the rim (or lip) to grind long, straight cutters. They maintain speed as they wear, as they do not lose diameter.

Tapered cups have some speed loss as wear occurs. They are used to increase the range of work on a universal tool grinder, since as they are tapered the wheel will fit more easily into position for grinding cutters attached to cutter blocks.

Dish (or saucer) wheels are saucer shaped and grind on the side of the rim. Again, they are shaped so as to enable them to fit into cutter teeth for grooving saws, etc.

Index